Przemysław Różewski, Emma Kusztina, Ryszard Tadeusiewicz, and Oleg Zaikin

Intelligent Open Learning Systems

T0181417

Intelligent Systems Reference Library, Volume 22

Editors-in-Chief

Prof. Janusz Kacprzyk
Systems Research Institute
Polish Academy of Sciences
ul. Newelska 6
01-447 Warsaw
Poland
E-mail: kacprzyk@ibspan.waw.pl

Prof. Lakhmi C. Jain
University of South Australia
Adelaide
Mawson Lakes Campus
South Australia 5095
Australia
E-mail: Lakhmi.jain@unisa.edu.au

Przemysław Różewski, Emma Kusztina,
Ryszard Tadeusiewicz, and Oleg Zaikin

Intelligent Open Learning Systems

Concepts, Models and Algorithms

 Springer

Dr. Przemysław Różewski
West Pomeranian University
of Technology in Szczecin
Faculty of Computer Science
and Information Systems
ul. Zolnierska 49, 71-210 Szczecin,
Poland
E-mail: prozewski@wi.zut.edu.pl

Prof. Emma Kusztina
West Pomeranian University
of Technology in Szczecin
Faculty of Computer Science
and Information Systems
ul. Zolnierska 49, 71-210 Szczecin,
Poland
E-mail: ekushtina@wi.zut.edu.pl

Prof. Ryszard Tadeusiewicz
AGH - University of Science
and Technology
al. Mickiewicza 30, 30-059 Krakow,
Poland
E-mail: rtad@agh.edu.pl

Prof. Oleg Zaikin
West Pomeranian University
of Technology in Szczecin
Faculty of Computer Science
and Information Systems
ul. Zolnierska 49, 71-210 Szczecin,
Poland
E-mail: ozaikine@wi.zut.edu.pl

ISBN 978-3-642-27078-9

ISBN 978-3-642-22667-0 (eBook)

DOI 10.1007/978-3-642-22667-0

Intelligent Systems Reference Library

ISSN 1868-4394

Typeset & Cover Design: Scientific Publishing Services Pvt. Ltd., Chennai, India.

Printed on acid-free paper

9 8 7 6 5 4 3 2 1

springer.com

Abbreviations

DLN   – Distance Learning Network
ICT   – Information and Communications Technology
IOLS   – Intelligent Open Learning System
LCMS   – Learning Content Management System
LMS   – Learning Management System
LO   – Learning Object
ODL   – Open and Distance Learning
OSLD   – Open System of Distance Learning
SCORM – Sharable Content Object Reference Model

Foreword

The book addresses the problem known in the literature as Open and Distance Learning (ODL). ODL describes new concepts of the process of learning-teaching organization. ODL is becoming very popular at universities that offer education via online and/or distance learning, both in synchronous or asynchronous modes. More precisely, the book addresses Intelligent Open Learning Systems (IOLS), i.e., the systems where traditional methods of online teaching are enhanced through the use of artificial intelligence methods. Taking this approach helps to achieve the desired teaching goals and greatly improves the quality of student learning. In the book, the IOLS concept is used in the sense of an information system for learning process management. Introduction of the idea of social and information system organization, such as ODL, requires deep analysis of its nature and structure. The complexity and scale of ODL is reflected in the IOLS management capabilities.

IOLS combines characteristics of the traditional understanding of the term Distance Learning and the new understanding of the term Open Learning. The open learning is focused on each individual student. Each and every student actively cooperates with the teacher in order to learn and "discover" new knowledge during their interactions. IOLS is designed to support this kind of collaborative work. It supports the social idea of global understanding and transparency of qualifications, and open access to educational resources. In the framework of new type of education process, such as the European Higher Education, every student in spite of his or her place of residence can participate at different national and/or university educational processes through the use of the Internet and similar technologies. Moreover, each student can achieve her or his educational goals based on personalized learning characteristics.

IOLS is a complex management system. Its many components have their own goals that determine their behavior. Moreover, these components are mutually related but can have conflicting goals. IOLS's complexity arises from the fact that it strives to satisfy and make use of many different stakeholders and resources: the students, subject matter experts, organizations, IT systems, telecommunication infrastructure, and information and knowledge resources. The success of IOLS implementation depends on the system's ability to efficiently manage its aggregate components. In order to automate parts of the IOLS the information system for its management should be based on an integrated model. Having such a model is a necessary condition for interpretation of the IOLS as one coherent system.

In the book the authors treat the IOLS as an information system for the open learning process management. The success in developing such a system has important research and practical implications. The practical side of the model is the development and implementation of the general methodology of the ODL so that it can be used by different people at different educational organizations. The ability to measure characteristics of the educational process is one of its main research outcomes. It is worth mentioning that the quality of the educational process is difficult to evaluate using traditional methods. The educational organization, like any other organization, must respond to changing environment. In case of educational systems the challenges come from using the ODL, which is the new approach to learning-teaching process organization. Each organization's success depends on having an efficient management system and the universities are no different. The every day practice of managing educational organizations needs to meet those new challenges. It needs to take into account the ODL concept that can be understood as a sequence starting from raw data, to information, to knowledge, and finally, to competence in the work place.

The authors analyze the ODL system based on the organization's mission and goals, market position and constrains, and include background knowledge understood as the basis of competence. Transformation of knowledge at any educational institution follows a specific workflow. The authors describe this workflow, its structure, and key characteristics. The presented information system enhances the learning-teaching process by making use of machine intelligence tools.

Every design of the information system should not only fulfill the requirements of mathematical verification but, ultimately, also meet the real-life validation. The authors validate the IOLS on the level of a student and an educational organization, while the real-life validation is performed using as the case study the AGH University of Science and Technology's Digital Student City. In contrast to existing information system for the learning-teaching process management, the IOLS is an example of a new class of information systems. Its main characteristic is dedication to the concept of openness. IOSL implementation affects social interactions due to different levels of its openness. This fact increases the importance of conceptualization phase of the IOLS design. Without having it it would not be possible to integrate different European educational institutions into a common widely-shared learning-teaching system.

The presented book expends traditional understanding of concepts, models and algorithms of the IOLS. The described results allow for creating new standard for this class of information systems. The standardization is performed at different levels and includes "soft" elements related to knowledge and competence. The book's contents reflect current state of knowledge and practical issues of e-learning. It covers concepts such as open information system, open access and content, competence management, ontology modeling, and tools for cooperation at the level of knowledge. The authors discussed these topics in the context of the European education system using real-world case studies and challenges they pose. For this reason I highly recommend the book to both researchers and practitioners of e-learning.

Krzysztof J. Cios
Virginia Commonwealth University
Richmond, U.S.A.

Contents

Part II: The Problem of Knowledge Modeling in Open and Distance Learning

Part III: Application of Open Learning Systems

Introduction

Motivation

Development of technology and civilization progress lead to necessity of general (e.g., spread on whole population) and permanent learning. In Information Society knowledge will be the top value and every members of such Society must continuously increase possessed knowledge. Life Long Learning (LLL) is no longer theoretical postulate – in fact it is actual imperative, very important for economical and social development of contemporary countries and societies as well as their civilization and cultural promotion.

Life Long Learning is not achievable on the base of traditional school model. Organizations like primary and secondary schools, colleges, universities etc. are off course still necessary, but are definitely not enough. For effective spreading of information, for sowing of knowledge, for growing up wisdom – entirely new methods, institutions and tools are necessary. Fortunately development of the Information and Communications Technology (ICT) gives now absolutely fantastic possibilities related to so called distance learning, internet-based learning, computer aided learning and many others methods, technologies and tools short named altogether e-learning.

E-learning is now quite widespread technology used in many schools for better, cheaper and more intensive teaching and learning procedures. But what we need in fact now is Open e-Learning System, because only open resources can be effectively used for achieving civilization oriented goals in the whole society. In presented book the Open e-Learning Systems will be described, discussed, proposed and evaluated. We concentrate on the most important and most modern type of Open e-Learning Systems, e.g., we will talking about Intelligent Open e-Learning Systems. Artificial Intelligence (AI) is very important and very useful element of every e-Learning Systems, because learning process is much more effective if is planed, organized, performed and assessed in intelligent way. This AI factor is especially need in Open e-Learning Systems, because such systems must be able effectively cooperate with many different learning people – exactly as it is planned in LLL politics. Therefore thinking about the future e-Learning system we must develop Open e-Learning Systems, improving their concept, perfecting their models, developing their algorithms and spreading their application.

Presented book contains all mentioned above elements. We describe basic concept of Intelligent Open e-Learning Systems, taking into account general evolution of learning and teaching systems as well as considering actual devices and possibilities of Computer and Information Technologies (CIT). In fact our approach is based on AI and CIT technologies and founded on system analysis principles. We discuss various models of Intelligent Open e-Learning Systems, derived from different theoretical frameworks: system science, cybernetic, computer technology, content based, educational tuned, psychological and pedagogical developed etc. Very important element of our proposition is related to competence based system designing, which is treated as a key for competence model building, intelligent teaching and learning system designing and system algorithms development.

Concepts, models, and algorithms as well as detail methods for achieving particular goals formulated for Intelligent Open e-Learning Systems discussed in the book are presented both theoretically and in context of practical applications. All applications of Intelligent Open e-Learning Systems discussed in the book are based on scientific results collected by the authors during over ten years practical research performed in the e-Learning area.

All elements described in the book: concept, models, algorithm, implemented systems and practical experience are devoted to one goal: to make Intelligent Open e-Learning Systems fundamental technology for development future Ubiquitous Open Life Long Learning System, which should be the answer for most challenging questions of XXI century civilization.

Book's Content

The book consists of ten chapters and is divided into three parts.

The first part concentrates on Open Learning System analysis. In the beginning of it, the social and educational meanings of the Open Learning System are discussed. One of the important outcomes is recognition of different openness contexts. The next chapter focuses on the distance learning environment. The new role of the teacher and the new requirements regarding the structure of didactic material characterize the new approach to learning environment design are discussed. In this chapter a cybernetic model of maintaining proper relations between the student, the teacher, and the computer, is proposed. Further on, the teaching-learning process is analyzed on the basis of the e-Quality project results. The authors took active participation in this project. The user's roles and activities are integrated in the student life-cycle. At the end of this part, the Open Learning Systems is analyzed using the system approach. As a result, the hierarchical structure and functional schema of Open Learning Systems are described.

The second part is focused on the problem of knowledge modeling in Open and Distance Systems. The first chapter covers the aspect of knowledge modeling basing on the ontology and the competence approaches. Afterwards, the Learning Object concept is discussed. In distance learning, ontology plays the role of the knowledge model. Authors propose an extended ontological model designed especially for open and distance learning. The next chapter covers the issues of designing and developing the learning objects and the corresponding knowledge repository. The last chapter of the second part talks about competence management in open systems. The competence set theory is used to build algorithms and methods for distance learning management.

The third part describes application of the Open Learning System. Firstly, the virtual laboratory for competence transfer is analyzed. The simulation experiment methodology serves as the basis for acquiring competencies in the virtual laboratory environment. Next, the application for competence acquiring is presented. The following chapter covers the issue of Distance Learning Network. A community-built system is the main paradigm for Distance Learning Network development. Furthermore, the motivation model integrating students and teachers on the social and knowledge levels, is described. At the end, a real-life application of the Open Learning System idea is presented. The AGH student city is a working example of the Open and Distance Learning System. Authors provide some thoughts about the future development of the open system idea basing on the AGH student city's evolution.

The content of the book was prepared according to the following pattern:

- Przemysław Różewski was responsible for chapters: 3, 4, 5, 6, 7, 8, 9
- Emma Kusztina was responsible for chapters: 1, 2, 3, 4, 6, 8
- Ryszard Tadeusiewicz was responsible for chapters: 1, 2, 10
- Oleg Zaikin was responsible for chapters: 2, 3, 4

Acknowledgement

The authors would like to acknowledge and extend gratitude to the following persons who have made creation of this book possible:

- Prof. Antoni Wiliński, Dean of the Faculty of Computer Science and Information Systems, West Pomeranian University of Technology in Szczecin, for his financial support, understanding and assistance.
- Dr. Katarzyna Sikora and Dr. Michael Maslowski for their language help and redaction.
- Regional Centre for Innovation and Technology Transfer in West Pomeranian University of Technology in Szczecin for assisting in the collection of data for research.

Przemysław Różewski,
Emma Kusztina,
Ryszard Tadeusiewicz,
Oleg Zaikin
Szczecin-Krakow, Poland, March 2011

Part I
System Analysis of the Open Learning Systems

Chapter 1
The Educational and Social Meaning of the Open and Distance Learning System: Polish Perspective

1.1 Introduction

Currently important changes are taking place in the European education system, caused by applying the system tool called Bologna Process. The Bologna Process is a social and organizational program that includes a set of activities concerning i.e. organization of the education process, ensuring transferability of the achieved results, creating a commonly accepted way of describing the obtained knowledge, aiming at creating the common European Higher Education Area. The main goals of the Bologna Process are: adapting the higher education system to the job market, creating an active European citizen behavior , transferring advanced domain knowledge and personal development of the educated person. The reason for taking on the spread-over-years task in the frames of the Bologna Process is the need to adapt the European education system to the challenges that occurred as a result of geopolitical and social changes, in order to create a modern information society. One of the challenges is to partially base the education system at the academic level on distance learning solutions.

This chapter concentrates on presenting the multidimensionalism of the Bologna Process. At the beginning the relation between information society and e-learning is presented. On one hand, the Bologna process is a result of the European Union's policy aimed at building a European education market. On the other hand, it responds to the idea of applying the concept of Open and Distance Learning, the goal of which is to open the education system for all social groups by providing proper computer tools and environment.

1.2 Information Society and Education

In different epochs of human history, different techniques played the leading role and were assigned crucial meaning. The 19th century was defined as the "age of steam", the beginning of the 20th century was called the "age of electricity", in the middle of the last century the "millenium of atom" was announced, etc. Therefore there is nothing unusual in the fact of connecting the end of the 20th century and the

P. Różewski et al.: Intelligent Open Learning Systems, ISRL 22, pp. 3–22.
springerlink.com   © Springer-Verlag Berlin Heidelberg 2011

beginning of the 21st with information techniques. However, as opposed to all the previously promoted "epochs" and "eras", computer science giving the tone to modernity has an incredibly strong influence on all the dimensions of functioning of entire societies and individual persons. This leads to a call for creating (with active support of all political, economic and social authorities) on the basis of this very technique – a new socio-economical formation of the so called Information Society.

This new term found its place in the consciousness of futurists and scientists and entered the arsenal of politicians, although it often raises objection and is also the subject of different discussions.

Critics of the idea of the Information Society can be easily understood. In their opinion talking about the new type of society only because most people use computers for many tasks seems like an overstatement. However, the skeptics are incorrect, as in essence the transformation brought by information techniques is so deep and significant that it supports such a concept (of the new socio-economical formation), which, what is more, once formulated, with time gains the strength of a self-fulfilling prophecy. Accepting this fact has long-reaching consequences for all domains of economy, society functioning and every-day life. This is due to the fact that the computer is an equally amazing and revolutionary tool regardless whether it changes the functioning of a big bank, allows for automation of car production, enables remote check of working hours of offices and makes it possible to do administration work without leaving home – or simply provides entertainment for a little child. What is more, in the history of civilization it is hard to find another example of a technique that would simultaneously transform all domains of human activity so deeply and radically as information technology.

In addition, this transformation occurred in an incredibly short period of time, what strengthened the feeling of its significance, as people immersed (not always voluntarily), pretty much from day to day, in the new, digital reality, usually lacked the time to fully adapt to these changes and assimilate the consequences. In effect, despite over half a century of dealing with computers, not everyone can fully use the advantages of this still avant-garde technique, not all people can plan their activities in such a way that they can be harmoniously adapted to the occurring changes, and we cannot fully protect ourselves from the threats that come with the dynamic development of information techniques.

1.2.1 Role of Computer Science in Education

In order to realize the direction and scale of the civilization processes we are currently a part of, as well as to show the common role of computer science in economy, technique and social life, a certain recapitulation of all these technical, civilizational, economical and cultural changes that occurred in the second half of the 20th century due to computer science, is needed. This allows for proper evaluation of all (incredibly complex) consequences of the situation in which the world found itself in the first decade of the 21st century. It is characteristic that at the moment when these changes took place, and we all participated in them in some way, nothing foretold that it is actually the announcement of a real revolution. Each consecutive invention and each following facilitation in the domain of information techniques were taken as something natural and obvious by most people. It is only today that looking from the current perspective we see the real scale of these ostensibly small and not really important events.

Discussing today how modern education should look like, to what degree and with what goal in mind should information systems be involved in the process of education on the way to distance learning, and also evaluating different techniques and methods of the distance learning itself, we need to realize that education is never a goal in itself. Education is always the mean to reaching the goal, and this goal is the intellectual, mental and moral shaping of people with a defined amount of knowledge, prone to a certain style of thinking and having specific value systems. Being at the doorstep of revolution that will create the future society as society of information, we need to be aware that the goal of our education is to prepare people to play certain roles and perform certain functions in the frames (structures) of this very information society.

This has wide-ranging consequences. The person functioning in the information society and the student taught for the information society, need to have a specific attitude towards information techniques, resulting from the fact that these techniques will then be used by them in all circumstances and for all purposes. Therefore, if the education process is also organized in a way that connects obtaining knowledge (regarding different topics) with using information techniques as educational tools, the chances that this way students will more successfully act as future citizens of information society will increase significantly. It is an additional and quite important argument speaking for a kinder approach to distance learning techniques, and for choosing them not only for the reason of economical savings or comfort of managing the didactic process, but also for the above-mentioned conditions.

During all epochs there was truth in the saying that we educate for the future. However, in the era of Information Society this saying becomes especially important and especially difficult for effective realisation, since we will only be able to educate well the future citizens of this future society if we can properly guess what this society will be like. In the meantime, no one really knows today what kind of a formation the information society will be and what will be the desired qualities and qualifications of its future citizens.

There is no doubt (since it is obvious even from just the name of this futuristic creation) that many characteristics and features of the information society will be determined by the development and progress of information techniques. The logic of development is roughly such that technology continuously opens new "niches", which are then filled by people with specific forms of activities and the social form connected to them. Thus if we could at least roughly estimate and imagine these new civilizational niches, we could foresee at least some of the future socio-economic conditions, what could be a good signpost for the development of education systems – in the area of the content being taught as well as in the area of desired forms and methods of teaching.

1.2.2 Information Society

Cheap, miniaturized and reliable computers lead to the situation today when equipment from the ICT group is applied genuinely everywhere and for everything. This fact resulted in discussions regarding the creation and development of the fore-mentioned Information Society [19] – and this keyword seems to most emphatically show the place and role of information techniques in the modern world.

The Information Society is not simply a community of people using computers at work, in public life, at home and in entertainment – it is essentially a different entity than the regular industrial (modern, as referred to by some humanists) society in which we were brought up and which we know well. In Information Society everything is different than in the earlier socio-economic systems, and therefore this slightly futuristic terminology is fully justified. This thread will be described in more detail further on.

It is said that as a result of the transformation towards the Information Systems quite soon in the structure of social division of tasks more people will be employed in creating, processing, distributing and analyzing different information than in producing material goods.

For people not used to the (currently) dominating role of industry this sounds strange and absurd – but it is true. Creation and processing, as well as distribution and analysis of different information will soon be the dominating form of economic activity, what is more – the most prestigious and profitable one. It is not a vision of futurists, it is a fact. Already today in the export of the USA more profit is gained from selling information (licenses, technologies, software, as well as movies, audio recordings, TV auditions etc.) than from selling industrial products, including weaponry.

It is worth noting that a schema is repeated here, a schema that took place in the past when due to the invention of different production machines and artificial energy sources (steam machine, afterwards electric energy) conditions were created for industrial mass-production and for creation of capitalism as a socio-economic formation providing resources for further development of industry and the model of economy based on it. It was a very radical and dramatic change. Through all the preceding centuries most people were employed in the production of food (in other words in agriculture, farming, hunting, gathering etc.), while other material goods were produced by sparse craftsmen, and the demand for their products was quite limited. In contrast, the scientific-technical revolution creating the civilizational prerequisites for the development of industry, and social revolutions (with the Great French Revolution and the American Declaration of Independence in the vanguard), released the potential for changes, the result of which is that in modern days industrial production is the synonym for modernity and wealth, while agricultural production is associated with poverty and backwardness.

Naturally the development of industry and weakening of the role of agriculture remain in a causal relationship, since it is the progress of mechanization and chemization of agriculture that made it possible for barely 5% of people in the developed countries of the world to provide food for entire societies. However, this has far-reaching social consequences: today the problem is not the lack of people to work with the land but their excess, and the vision of starvation caused by the existence of too many people in comparison to the productivity of the available land resources used for agriculture (Malthus' vision) was replaced with the perspective of over-production of food and connected to it (unfortunately) different forms of wastage.

Today other big changes are taking place, which Alvin Toffler called the "third wave". The progressing automation and robotization of production work transforms the industry of the 21st century in such a way that in order to produce an increasing number (and wider assortment) of material products, work of fewer people is required. Actually, from the technical point of view, completely automated plants producing material goods are possible to realize already today, although they are still not very popular due to economical reasons (human work tends to be cheaper) and social reasons (massive replacement of people with automates can lead to social unhappiness and political disturbances). Nevertheless, there is a noticeable and rapidly growing trend to achieve increasing industrial production with a decreasing number of people involved in the production processes – what is a global phenomenon and most probably has irreversible character. Simultaneously with the release of people from the production processes, the demand for different kinds of information increases. The natural consequence to both these facts is the involvement of people in the very processes described above, connected to creating and processing, as well as distribution and analysis of different information.

The Information Society that is the consequence of these changes will be characterized by the fact that many things that are now obtained in the real (physical) way will finally be somehow immersed in the so-called cyberspace, thus will undergo a kind of virtualization [13]. Below several attributes of a citizen of the Information Society were described. We treat it as a reflection of computer science of today, even though the future tense was used in the description. However, this vision – already partially realized in the most economically and technically developed countries, and undoubtedly appearing in the other countries soon – is not considered a prognosis (which would by its own nature be more far-sighted) but rather a description of the current situation.

Let us begin with the fact that due to the development and popularization of information techniques the model of performing work is already now changing (and it will change even more). The old concept of a workplace is disappearing, because the modern employee (producing information goods, according to the main principle of the Information Society) will be able to fulfil his/her tasks at any place, working at home, while traveling, outdoors or in any other place, thanks to the possibility of realizing the so-called tele-work, also known as e-work. If such a person gets sick, in the first instance he/she will receive help from a tele-medicine system, also called e-medicine [14]. If he/she wants to deal with administrative work – services of e-administration will be available [15]. When new qualifications or knowledge will be needed – he/she will be able to use e-learning methods [16]. A humanoid robot will help with house-work [17], and e-entertainment will be brought by interactive IT tools [18].

All these attributes are already present and available today, and the expected progress will mostly deal with their popularization. As can be concluded from the presented remarks – Information Society is already somehow present in Poland of the second decade of the 21st century, and its development seems unavoidable.

Each domain of technology and economy will pretty much have only two possibilities: adapt to the occurring changes, by undergoing a far-stretched computerization, or be a part of the increasing marginalization [6]. Therefore also education processes need to be connected and harmonized with the concept of Information Society. To realize this task it will be beneficial to attempt foreseeing in which direction information techniques will develop. We will thus try to give, at the end, several remarks that will constitute for the attempt to forecast the development of Computer Science and Informatization.

1.3   Trends in the Development of the Education System

Joining the European Union by Poland created the need to adjust the Polish education system to the European system like other European Union countries. Because the European system is now undergoing a transformation according to the Bologna Process, knowledge is needed about the goal form of it. The performed analysis is enriched with an additional dimension of distance learning as an important component of education system being created.

1.3.1   Analysis of European Union's Educational Policy

In its actions, the European Union gives proof of caring about the educational issues. The ambitious plan of building an information society – knowledge society – must be based on solid fundamentals represented by versatile educated citizens. Poland, like other European countries, is moving towards information society, as was shown in [2]. Defining the direction of developing the European education system towards solutions based on distance learning began already at an early stage of constructing the European Union. Resolutions of Treaty of Maastricht (7th of February, 1992) were a milestone in this direction. Apart from ruling for a common currency of the Union, in article 126 the issue of education (including distance learning) was considered for the first time. Article 126 says, i.e. „(…) The Community shall contribute to the development of quality education by encouraging co-operation between Member States (…)Community action shall be aimed: (…) – encouraging the development of distance education".

The process of building systems and structures of distance learning in European Union, the result of which is the presented above formal resolution, has started long before. The analysis performed in [20] shows the evolution of distance learning systems that took place in European countries. Consecutive actions undertaken by UE gradually reorganized the approach and the applied perspective. In the beginning, distance learning in Europe was represented by a set of autonomously functioning universities, each of which independently offered their services (e.g. *Open University*, active in England since 1969, or *Fernuniversitaet*, functioning in Germany since 1974). At the next stage, organizations and institutions teaching over distance started joining into consortia according to the model oriented on specific areas of Europe or vocational groups. Specialists and decision makers of European Union, in their successively published working materials and official documents, were systematically building the structure of the mission that should be fulfilled by systems and organization s of distance learning.

The most important task, placed by the European Union in front of the system solution to the issue of education (based on the distance learning technology), was equaling the educational chances of all citizens of European Union. The education process should promote and develop the European cultural heritage. The bridge connecting citizens of the Union in one organism of knowledge civilization will be built through creating a mutually understanding society. Distance learning systems supported by the Union's programs are aimed at creating a level of general knowledge of European Union's languages, which would allow easy communication at all levels: neighborhood-related, local, national. The distance learning systems have special meaning in educating the migrating work-force. On one hand these systems allow for quick acclimatization of the migrating employees in the new environment (e.g. through getting to know a certain language), and on the other hand ensure the possibility of increasing or changing qualifications of the employee depending on the market requirements.

The practical results of the accepted resolutions are the research and educational programs started as a part of the European Union's educational policy. Each program is designed in such a way as to activate the defined areas of common educational activities. Characteristics of the most important programs of European Union connected to education, especially distance learning, can be found on appropriate WebPages of the Union's service.

European Union's programs include the context of each participating country. In case of Poland the programs are designed bearing in mind the governmental changes that occurred in Central and Eastern Europe. As was shown in [5], after the changes that happened in the 90', universities of the earlier communist countries found themselves in a completely new situation. They obtained an unknown before level of autonomy and freedom of decisions. Actions were performed to renovate the missions, syllabuses, organization structure. Methods of financing scientific, research and educational activity were redefined. Increasing number of students caused the creation of a resilient sector of educational services offered by non-public universities. New conditions resulted in a change in the way of functioning of each university. In its activities, European Union aims at connecting all European higher education organizations in one common education area, built on the basis of technological solutions of information society, i.e. with the help of distance learning systems.

1.3.2 Bologna Process

The Bologna Declaration was signed in 1999 in order to integrate the ideas and activities of European countries for the development of education in Europe. The main goals of the Bologna Process are expressed through the following points [7]:

- preparing graduates for the labor market needs;
- preparing to be an active citizen in a democratic (European) society;
- development and support of advanced knowledge (society and economy of knowledge);
- personal development of the educated people.

The final idea is creating a European Higher Education Area (EHEA), with the help of different system tools, including the Bologna Process. The Bologna Declaration does not close the possibility of participation for other countries that wish to commonly develop in the frames of the ideas supported by it. According to the declaration, it is essential to increase the level of standardization of administration procedures, i.e. in the process of developing syllabuses and courses of the three-cycle education and introducing a credit-based evaluation of students' achievements. The presented changes will support promoting mobility of the research and educational staff of a university, and, what is most important – mobility of students.

The Bologna Process can be seen as a special kind of socio-political movement aimed at transforming the education system of European countries into a form of a common area of offering educational services on the basis of adapted tools and methods (see tab. 1.1.).

One of the more important aspects of the described Bologna Process is developing a uniform system of describing knowledge that is passed to the students in the frames of the consecutive higher education cycles. In the European Higher Education Area each student can individually form his/her education path basing, among others, on the description of competences offered through courses at different universities [7]. The form of the description of obtained competences allows also for recognition of the current state of knowledge of the student by a different university or an employer. Currently work on this issue is being carried out in the frames of the Bologna Working Group on Qualifications Frameworks. The concept of the Bologna Process assumes creating a pan-European system of competences/qualifications. It is planned to create two complementary systems of competences/qualifications. The general one, called Dublin Descriptors (see the work of the Joint Quality Initiative project: http://www.jointquality.org/, and fig. 1.1), concentrates on competences such as e.g. communication skills, understanding, developing judgments. The detailed one (see results of the Tuning Education Structures in Europe project), is prepared for each domain, e.g. mathematics, chemistry. The detailed descriptors create the final form to which education standards of teaching individual study programs and specializations will be adapted. This means that instead of describing the past knowledge, expected competences of a student will be characterized for each study program.

Table 1.1 Historical outline of the development of the Bologna Process (based on [7])

Stages	Resolutions and activities
Bologna Declaration (19th June 1999)	Activities: Diploma Supplement; Three learning cycles (bachelor, master, doctorate); ECTS - European Credit Transfer System. Aims: Introducing mutual understanding and comparable grades system; Promoting mobility of students, researchers, academic and administration staff; Promoting European cooperation for enhancing the quality of higher education.
Prague meeting (18-19th May 2001)	Activities: Commitment of higher education institutions to the qualification framework. Aims: Increasing the importance of lifelong learning; Increasing the attractiveness of European Higher Education Area for students from Europe as well as from other parts of the world.
Berlin meeting (18-19th September 2003)	Activities: Quality validation framework; European Network for Quality Assurance - ENQA. Aims: Developing and implementing quality control procedures
Bergen meeting (19-20th May 2005)	Activities: Creating European Register of Accreditation Agencies. Aims: Adapting the European Quality framework based on the ENQA Works: Adapting the European Framework for Qualifications.
London meeting (17-18th May 2007)	Activities: Add the social dimension aspect to the European Higher Education Area (EHEA) key activities. Works: Evaluation of mobility, degree structure, recognition, qualifications frameworks.
Leuven/Louvain-la-Neuve (28-29th April 2009)	Aims: New higher education priorities: social dimension: equitable access and completion; lifelong learning; employability Works: Multidimensional transparency tools for data collection.

Cycle	Knowledge and understanding	Applying knowledge and understanding	Making judgements	Communication	Learning skills	
Bachelor (1) 180-240 ECTS credits	[Is] supported by advanced text books [with] some aspects informed by knowledge at the forefront of their field of study	[through] devising and sustaining arguments	[involves] gathering and interpreting relevant data ..	[of] information, ideas, problems and solutions ..	have developed those skills needed to study further with a high level of autonomy ..	The cognitive model of competence acquire
Master (2) 90-120 ECTS credits	provides a basis or opportunity for originality in developing or applying ideas often in a research context ..	[through] problem solving abilities [applied] in new or unfamiliar environments within broader (or multidisciplinary) contexts ..	[demonstrates] the ability to integrate knowledge and handle complexity, and formulate judgements with incomplete data ..	[of] their conclusions and the underpinning knowledge and rationale (restricted scope) to specialist and non-specialist audiences (monologue) ..	study in a manner that may be largely self-directed or autonomous..	
Doctorate (3)	[includes] a systematic understanding of their field of study and mastery of the methods of research associated with that field..	[is demonstrated by the] ability to conceive, design, implement and adapt a substantial process of research with scholarly integrity ..	[requires being] capable of critical analysis, evaluation and synthesis of new and complex ideas..	with their peers, the larger scholarly community and with society in general (dialogue) about their areas of expertise (broad scope)..	expected to be able to promote, within academic and professional contexts, technological, social or cultural advancement ..	

Inherent pattern during the competence acquire process

Fig. 1.1 Schema of general Dublin descriptors (introduced by the Joint Quality Initiative) for three cycles of education (source [7])

1.3.3 Educational Strategy of European Union

Let us analyze the planned directions of development in the educational strategy of European Union for years 2007 – 2013 on the basis of [1,4], taking into account the place and new role of distance learning systems. The meeting of the European Union experts in Lisbon in 2000 set a goal for the discussed European programs – until 2010 the citizens of Europe should create society based on knowledge. Performing such an ambitious task requires i.e. increasing social integrity by reducing disproportions between individual regions of Europe. The main tool allowing that is education. In education programs accents are placed on activating cooperation and mobility. Solutions at the level of distance learning, in the opinion of European Union specialists, will allow fulfilling requirements placed before the society based on knowledge economy, and are also a realistic solution for problems of mobility of the Union's citizens.

The direction taken by the European Union is only to a certain extent in accordance with the view on educational future present in America and on other continents. In discussions about the future of education from the point of view of the United States, [11] predicts that in the near future (around 2030) academic education, especially due to the increasing cost, will take place in two types of institutions:

- Experience Camps – on the basis of federal funds and donations from companies, centers are created, in which a limited number of students (<1000) conducts experiments and projects. In the time span of around two years students acquire knowledge and abilities in a certain, narrow domain. The discussed solution is not considered as a general solution for everyone, it is directed only at the most talented individuals.
- Advanced Learning Networks – commonly available educational solutions for less wealthy, talented students.

Both the European Union's authorities and individual member countries see the future of education (especially academic education) in the same way. Distance learning is one of the most successful solutions for the dilemmas of mobility and uniform level of education in the Union. Distance learning systems are the natural platform for cooperation between different universities. In Poland increasing support for such approach can be seen on the example of many initiatives both at the level of entities (a number of local initiatives consisting of creating distance learning servers with didactic materials for several courses), and at the level of ministry (projects for developing didactic materials for distance learning of basic academic education programs in Poland).

1.4 The Concept of Open and Distance Learning

The main idea of Open and Distance Learning (ODL) is connected to fulfilling the widely understood mission of Information Society. It covers all activities directly or indirectly connected to the process of distance learning of different people, students and pupils, and provides appropriate infrastructure and legislation. Distance learning systems are interpreted as the technological basis for the more general concept of ODL. The concept of ODL will be discussed on the basis of [8].

In the report developed by [12] each aspect of ODL was thoroughly analyzed. The distance aspect defines the educational situation in which the student is far away in space from the didactic materials and the different participants of the learning process. Communication with the system and other participants occurs only on the basis of the prepared computer environment. Such an approach is classified as learning- teaching in an asynchronous mode, and is the opposite of synchronous learning based on online lectures. In ODL on the basis of hardware and software solutions an individualized, virtual learning space is built. Basing on the methods of artificial intelligence, educational sequences are created, and are, on the basis of certain pedagogical methodologies (in the best case constructivism), passed to the student.

The openness aspect of the ODL process is visible in the strategy and policy that underlies the approach. Each user is to have the possibility to freely choose the material he/she will learn, and the place of study. An important aspect of openness is the message behind the entire idea of distance learning – distance learning systems make the same (usually high quality) didactic material available for each participant of the learning- teaching process. Such an approach allows realising the idea of equal educational chances for everyone, what is one of the main factors of developing ODL in Europe.

An important aspect, that influences an ODL-based education system, is the change of the paradigm of education, which evolves from the traditional form to the form of distance learning based on the internet. Introducing this new system, represented by the idea of ODL [12], will ultimately separate the process of gaining education from the place where the providers of the learning content are situated (meaning the university base) and from the place where the student is present at a given moment [6]. Except for special situations, for example connected to the process of certification, this stands for the possibility of breaking up with the three main uniformity rules of traditional education: uniformity of place, action and time. As the challenges connected to the mobility of European citizens require education to be organized in such a way that the learners and teachers could meet in the information space (so-called cyberspace), the uniformity of place does not have to occur [19]. Professors can provide the ODL system with a learning content in place X and their students can use this and other contents in places $Y_1, Y_2, ..., , Y_n$. Moreover, no two places need to be identical and a functioning connection of the professor with each of the students can still be fully guaranteed. Also, without much hustle, the conditions of a constant, close communication between the students can be ensured, even if each of these students is in a different place or traveling at the moment.

Similarly, it is not necessary to maintain the uniformity of time. In the traditional learning process the lecturer and all the students listening to the lecture have to perform actions connected to their roles exactly at the same time. Students that came late to a lecture irrevocably lost some part of it. It does not have to be this way anymore. The lecturer can supply the ODL system at the moment that is most convenient for him, the student can derive knowledge from the ODL system when he is best prepared physically, mentally and organizationally, and all interactions (greatly desired in the learning process) can occur completely asynchronously [3].

1.4.1 Aspects of the Open and Distance Learning in the Scale of European Union

In order to fulfill the concept of Open and Distance Learning following elements should be discussed in the context of the learning systems:

- Social aspect, the widening of which results mainly from two causes: firstly, the area of the system's functioning is no longer limited to a certain group of people but to an entire community, in the frames of the weakening meaning of geographical borders; secondly, quick desactualisation of domain knowledge and changes in the working conditions result in a social need for life-long learning.
- Cultural aspect, which covers managing the process of diffusion of cultures and national traditions with the need of creating European values, while maintaining status quo for each ethnic and social group.
- Technological aspect covering the issues of assessing the scientific, technical and organizational possibilities of supporting the functioning of ODL as a system for managing the planning, preparing and providing of a personalized educational service.
- Broadening of the economical aspect results from the need to increase the required level of complexity and universality of hardware and software resources. This activity increases the cost of ODL, what is contradictory to the idea of fulfilling social requirements that assume lowering costs of education with a simultaneous increase in its quality. Finding a compromise becomes the main economical problem.
- Arising of the political aspect is caused by the fact that ODL, which crosses national borders with its scope, requires direct involvement and participation of not only EU bodies, but also of the governments of all involved countries.

The need to include humanistic values, personalization of educational services and scale of popularization, qualify ODL as a social system which requires formulation and appropriate interpretation of certain philosophical values, i.e.: emergency, synergy, holism and isomorphism.

In general, emergency is the phenomenon of creating a new, previously not mentioned value as a result of the system functioning in a new environment. The new value appears as a result of cooperation between the elements of the system, the characteristics and parameters of which are already defined. This new value becomes a characteristic of the system as a whole. In regard to ODL, what should become such a new value is the universal for the entire education system form of knowledge representation as a shared resource, and the semantic model of describing the system of managing the learning- teaching process, reflecting not only its flow, but also the content and quality of knowledge obtained by the student at each stage of education.

Synergy is the effect of strengthening each part and the entire system as a result of their cooperation. In the context of open learning systems, apart from traditional planning, organization and administration of the education process, the issues of volume, depth and structure of knowledge become the object of management. Synergy in this context is expressed through interaction and cooperation, with a certain level of applied knowledge. The summary competence obtained by all participants during the execution of the education process will express the increase of the intellectual capital of the educational organization working according to the ODL principles.

By holism we denote the evolutionary process of creating new entities, considering that the value of the whole has preponderance over the sum of its parts. In that sense, the process of gathering and assimilation of the shared knowledge resource in the frames of big educational organizations of the world obtains the effect of holism. Knowledge resource can be presented as a multi-level structure, where the top levels correspond with the higher, and lower levels with the lower, level of abstraction and generality.

We are dealing with isomorphism when the form is system-creating, forming the structure, with greatly different content filling it. Isomorphism is the measure for the appropriateness of structures and functions of objects. In reality, the phenomenon of complete isomorphism does not exist. However, in cognitive processes, where we use abstractions, isomorphism allows explaining the sense of a concept through structuring of its definition. In regard to ODL, isomorphism is the effect of maintaining the structure and functionality of the system at different scales of implementation. Scalability parameters in this case are often quite different: quantitative parameters of the educational organization, domain knowledge and cognitive styles of teaching.

The mentioned philosophical rules are the essence of maintaining an open learning system (as a uniform object), what defines the need to include them in the model of managing every educational organization involved in the idea of ODL.

1.5 Analysis of the Openness Issue in Distance Learning Systems

In the learning systems, the concept of openness can be considered regarding different aspects. Changes that occur in educational policy of the new European Union members require a more specific outline, in a way that allows for defining a uniform approach to the concept of openness.

An example of an open system close to the problems of education is the natural language. Adding new words as a result of different causes (e.g. new technologies, expanding one's own environment of life) does not disturb the main function of the language which is ensuring communication in situations previously undefined. The main requirement for maintaining this functionality on an unspecified interval of time is the possibility to use the language to formulate new thoughts on the basis of the already stated ones. Following the ontology of thought written in a verbal and symbolic way may serve as a basis for modeling a common resource of knowledge for different groups of people.

1.5.1 Openness of the Education System in Social Aspect

Openness at the social level means the possibility to flexibly adapt the education system to changing social requirements regarding educational services, and providing each citizen with the right to increase their competences through the use of the education system in any mode.

The main area of functioning of the education system is society as a whole, and each receiver of the educational services individually. The leading idea of open learning is connected to fulfilling the widely understood educational mission of society. The mission covers all activities directly or indirectly linked to the process of distance learning of different groups (including handicapped) of students and pupils, providing appropriate infrastructure and legislation.

The higher the cultural level of a country, the more demanding the bylaw of education (scope of the program, time and mode of learning, certification conditions). One of the goals of standardization is to increase the average intelligence resource level of the society of countries included in the Bologna Process. Introducing standards does not aim at abolition of cultural and social specificity of individual countries, but guarantees a successful exchange of students, study programs, didactic material, or conducting internships abroad.

In the social context we have to talk about mutual openness of the human and the society for evolutionary development of humanistic and social values. For a person this means that he/she should be prepared for continuous education throughout his/her entire life. For the environment this means continuous adjustment of the infrastructure, legislations and social security.

1.5.2 Openness of the Learning System in the IT Aspect

Distance learning systems are the technological basis for development of the open learning concept. In the scope of computer systems, an open system is a piece of software with flexible structure, where the relations between modules are described in a formal language, and it is possible to add new modules and set (in specified range) the parameters of functioning of each module, according to the implementation conditions. The number and content of the modules to a great extent depends on the goal and criterion of using the system by the distance learning organization.

Analysis of individual cases of implementation of distance learning systems shows that the distance learning system is understood as a computer (IT) tool, which is in control of several, differently educed processes (i.e. managing student accounts, managing courses, facilitating communication between participants of the education process, reporting, doing statistics, supporting education in the frames of a single topic/course, using simulators for obtaining different vocational abilities). A strict list of these processes, their classification, executing rules for distance learning systems, was not developed yet.

Analysis of trends in the openness of education systems regarding the computer aspect shows that at the moment solving this problem depends less on hardware, network issues, and more on the possibilities of formal representation and manipulation of knowledge presented in the education process.

1.5.3 Openness of Distance Learning Corporate Networks

Each university included in ODL can be treated as a corporation working as a distributed organization. Corporate networks are usually used to create a uniform network environment in which the following functions are performed: planning, managing production, administration, logistics, managing clients and staff.

The market environment and the competition force the university to obtain characteristics of an enterprise and apply its own methods for maintaining the position on the market. Methods of using network resources play an important role in this case. One of the main characteristics of an educational enterprise is the fact that it is oriented on serving a random stream of orders for educational services on the basis of intangible production [22]. Each workplace is located in the network, the product/half-product is created as a result of cooperation between all participants of the education process, and it is continuously sent over the network. The change in the role of the teacher, the necessity of his/her cooperation with the new body of specialists (e.g. expert, knowledge engineer), the constantly changing network environment, require continuous uplifting of the qualifications of the staff. This process is conducted more and more often through the methods of distance learning. Therefore, an educational organization additionally acquires the characteristics of a learning organization [9,10].

The mentioned above characteristics of educational organization working in ODL conditions show that this organization uses the features of the corporate network in a much wider sense.

The open corporate network is a computer-based way of realizing cooperation, and information and knowledge exchange between members of the organization and the distributed structure. The main requirements of the distributed organization towards the corporate network are the following:

-   scalability – possibility to change, at any moment, the number of clients and the functional software packages;
-   high efficiency – ability to process high amounts of information; guarantees quick performance of complicated applications based on the client-server interface or Grid;
-   elasticity – possibility of automatic or semi-automatic adjustment of the corporate network to often changing conditions of using the network within the distributed organization.

The corporate approach allows for connecting the requirements of openness of systems in the social sense with the openness in the computer sense. Planning and development of the corporate network requires considering this problem in the context of managing the entire educational enterprise, based on intangible production.

1.5.4 Openness of the Student Life-Cycle

The student's life cycle is a process of cooperation between the student and the university, therefore the issue of openness should be looked at from both points of view. Open student life cycle means that each student has the possibility to choose the learning program on the basis of certified education offers of European universities, according to their own point of view, but in accordance with the defined rules. The basis for such an approach is the European Credit Transfer System (ECTS) [7].

The main rule that allows for choosing individual programs of studies, in an ideal situation, should be a universal for all universities taxonomy: specialization – profile – subject. In the frames of this taxonomy, a relationship of many – to many should be defined between the specialization and the subject, what means that one subject can be a part of study programs for different specializations, and different specializations can have common groups of subjects.

The next rule assumes that regardless of what education program the university offers for teaching a certain specialization, each subject passed within it has a weighted indicator according to which the student obtains a certain amount of commonly accepted points (ECTS points).

Another rule considers defining a minimal sum of points that each student has to obtain during studies following his/her own education path. A given sum of points is the basis for awarding a certificate (diploma) of the specialization.

The last rule assumes there is a possibility to compose one's own study program by the student in such a way that he/she will pass individual blocks of the program at different universities. The document facilitating recognition of the points obtained by the student will be the diploma supplement.

The student life cycle in the traditional learning conditions is a process of obtaining knowledge according to a specified education program that clearly defines the content and schedule of this process at one university. Distance learning allows for a freer schedule, performed also at several universities. The student's strategy will then boil down to simply choosing the university that guarantees achieving the appropriate certificate – the student's object of interest. After choosing a university, the student is included in a deterministic student life cycle. From the point of view of the university, the process of servicing all students is a deterministic one, and teaching certain subject to different groups of students can happen in parallel. The process of carrying out the defined study program can be described by the university in the form of a Gantt chart.

In the ODL conditions, the student's life cycle is a stochastic process. The student, while maintaining the same object of interest, can choose subject (courses) at different universities. The student is able to check at which universities the proposed study program guarantees obtaining the proper certificate. He/she then independently chooses at which universities to realize his/her won student life cycle. In other words, students move between universities realizing a personalized education path. The structure of the student's life cycle is constantly open. After passing a certain block of subjects at a chosen university, the student again faces the situation of choice as to where to continue the education in the future. In this case, on a certain horizon of planning each university will be dealing with a random number of students at each subject included in the learning program. In the described conditions, the process of teaching students becomes distributed in the common education space. The education space can be presented in the form of a mass-servicing network, where each node represents a university. Functioning of the so-described process can be presented using a stochastic model of the mass-servicing network (Jackson's model). The education process in each of the nodes of the entire network can be interpreted as pipeline production. Table 1.2. shows the comparison of functioning of distance learning (DL) systems and open learning (ODL) systems from the point of view of servicing students.

Table 1.2 Comparison of functioning of Distance Learning systems and Open and Distance Learning systems from the point of view of student processing (source [8])

Object of investigation	Distance learning (single institution)	Open and Distance Learning (institutions' network)
1. Learning curriculum for student of j specialization (P_j^{ST})	The learning curriculum is compatible with university U curriculum j $$P_j^{ST} = \delta_j^U \left(P_j \right)$$ δ_j^U — the probability that student selects the curriculum j from the university U	The student's curriculum consist of the courses from different universities $$P_j^{ST} =$$ $$(\delta_{j1}^{U_1}(p_{j1}) / \delta_{j2}^{U_2}(p_{j2}) / ... \delta_{jn}^{U_k}(p_{jn}))$$ $\delta_j^{U_k}$ – The probability of student selection of specialization j on the university U $p_{j1}, ... p_{jn}$ - curses of j specialization $U_1, U_2, ... U_k$ - universities caring the j specialization
2. Infrastructure for learning process	Virtual organization,	Distributed ODL organization (characterized by the stochastic student's flow)
3. Student's servicing model	Queuing system (M/M//N/$^\infty$)	Open queuing network with stochastic transmission (Jackson network)

Figure 1.2 presents how ODL works, with the market of required specializations and the potential contingent of students found outside of the educational organization, creating its surrounding, to which it should comply. Adjusting to the market requirements in the organization is conducted in two stages: in the beginning through development of appropriate profiles, which the organization offers on the market of educational services, followed by the development of a study programs according to which the order of a student interested in the offer can be processed. In order to maintain the position on the educational services market, the educational organization should constantly ensure openness of the student's life cycle.

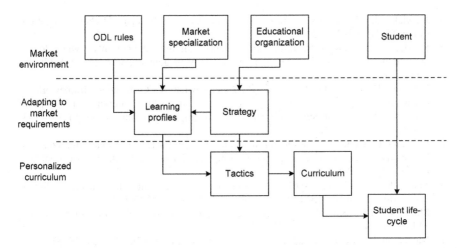

Fig. 1.2 Adapting the educational organization to market requirements

1.6 Conclusions

The presented concept of distance learning, compliant with the Bologna Process, undergoes constant evolution caused by the difficulties in adapting the presented idea to the changing conditions of the education market being created. The result of this is the organization of regular meetings (until now the meetings happened in: Bologna 1999, Prague 2001, Berlin 2005, Bergen 2005, London 2007, Leuven 2009) of education ministers of the countries that participate in the Bologna Process, in order to introduce required corrections and specify the strategy for the next period of time.

ODL defines the goal structure, which represents coexistence of the education market and the requirements and social mission behind the education process. In the future some form should be found that will include in its dimension commercial educational solutions together with state education. The vision of European education is additionally supplemented by distance learning, which will become a significant complacent of education conducted in the traditional mode.

References

1. Bednarska-Chłopaś, A.: European educational standards. Computer Science: Informatyka Teoretyczna i Stosowana 3(4), 225–236 (2003) (in Polish)
2. Cellary, W.: Poland on the way to the global Information Society. United Nations Development Programme, report (2002) (in Polish)
3. Chrzaszcz, A., Gas, P., Kusiak, J., Tadeusiewicz, R.: Learners – Teachers on-line mutual support. In: Szücs, A., Lngeborg (eds.) Proceedings of the EDEN Conference, Budapest, June 16-19, pp. 522–527 (2004)
4. COM 156: The New Generations of Community Education and Training Programmers after 2006. Communication Form the Commission of the European Communities, Brussels (2004)

5. Jóźwiak, J.: Perspectives from East and Central Europe. Higher Education Policy 15(3), 263–276 (2002)
6. Kicki, J., Tadeusiewicz, R.: Information science in mining and not only - where do we are and where do we aim? Mineral Resources Management 23(4), 111–135 (2007) (in Polish)
7. Kusztina, E., Zaikin, O., Różewski, P., Tadeusiewicz, R.: Competency framework in Open and Distance Learning. In: Proceedings of the 12th Conference of European University Information Systems EUNIS 2006, Tartu, Estonia, June 28-30, pp. 186–193 (2006)
8. Kusztina, E.: Conception of Open Information System for Distance Learning, Szczecin University of Technology, Faculty of Computer Science and Information Systems, book (2006) (in Polish)
9. Mack, R., Ravin, Y., Byrd, R.J.: Knowledge portals and the emerging digital knowledge workplace. IBM Systems Journal 40(4), 925–955 (2001)
10. Marwick, A.D.: Knowledge management technology. IBM Systems Journal 40(4), 814–830 (2001)
11. Nicholson, P.: Higher education in the year 2030. Futures 30(7), 725–729 (1998)
12. Patru, M., Khvilon, E. (eds.): Open and distance learning: trends, policy and strategy considerations. UNESCO, ED.2003/WS/50 (2002)
13. Tadeusiewicz, R.: Modern financial administration as a source of new scientific problems and application challenges for automatics and computer science. In: Gilowska, Z., Tadeusiewicz, R., Tchórzewski, J. (eds.) Krajowa Administracja Skarbowa Tom 3 - Nowoczesna Administracja Skarbowa, Difin, Warszawa, pp. 11–29 (2007) (in Polish)
14. Tadeusiewicz, R.: System approach to some telemedicine problems. In: Kulczycki, P., Hryniewicz, O., Kacprzyk, J. (eds.) Techniki Informacyjne W badaniach Systemowych, pp. 341–360. WNT, Warszawa (2007) (in Polish)
15. Tadeusiewicz, R.: The e-government development as a component of the information society. In: Siwik, A. (ed.) Od społeczeństwa Industrialnego Do Społeczeństwa Informacyjnego, pp. 391–404. UWND AGH, Kraków (2007) (in Polish)
16. Tadeusiewicz, R.: Which of the e-education models: robot or megaphone? In: Morbitzer, J. (ed.) Komputer w edukacji, Pracownia Technologii Nauczania AP, Kraków, pp. 253–259 (2007)
17. Tadeusiewicz, R.: Do we await androids? In: Proceedings of the conference Współczesne Problemy Inżynierii Mechanicznej, WIMiR, vol. 34, pp. 83–97 (2007) (in Polish)
18. Tadeusiewicz, R.: Varicolored computer science. Pomiary, Automatyka, Kontrola 5, 4–5 (2007) (in Polish)
19. Tadeusiewicz, R.: Internet Community. Shaker Verlag GmbH, Aachen (2005)
20. Tait, A.: Open and Distance Learning Policy in the European Union 1985-1995. Higher Education Policy 9(3), 221–238 (1996)
21. Zaikin, O., Kusztina, E., Rozewski, P., Malachowski, B., Tadeusiewicz, R., Kusiak, J.: Polish experience in the didactical materials creation: the student involved in the learning/teaching process. In: Mahnic, V., Vilfan, B. (eds.) IT Innovation in Changing World, Proceedings of the 10th International Conference of EUNIS, pp. 428–433. University of Ljubljana (2004)
22. Zaikin, O.: Queuing Modelling Of Supply Chai In Intelligent Production. Informa, Poland, Szczecin (2002)

Chapter 2
Distance Learning Environment

2.1 Introduction

Modern technologies allowed for breaking the immemorial paradigm of educating. It was mainly due to internet that the phenomenon of distance learning emerged [6]. Earlier, learning relied on creating an integral relationship of the teacher and the student. Such conjugation allows for creating an effective platform of knowledge exchange. The existence of human brain both on one as well as on the other side of the common education space provides a way of communication limited only by natural language. Currently this relationship can be simulated or reflected in the internet environment.

In the beginning of the considerations we need to define the semantics of the concept of distance learning. Distance learning over the years evolved in two main directions: synchronous and asynchronous learning. The first one is characterized by uniformity of time, students and teachers are participants of one learning session, separated by space they communicate with each other using modern computer technologies. The considerations will, however, be limited to the area of asynchronous learning, due to the economic advantages of this mode. Furthermore, asynchronous learning is a more important research problem, since its main idea assumes reflecting the process of learning- teaching, which until now took place in the minds of students and teachers.

A rich set of definitions of distance learning in asynchronous mode can be found in [23]. The initial definition of the term was proposed by [22]: „the term of distance learning was applied in many training methods; however, its basic definition assumes division in space and division in time regarding the contact between the teacher and the student". The asynchronous aspect of division in time between the participants of the education process is discussed in more detail in [6] with consideration of the paradigm of asynchronous learning networks (ALN). Following [6] we can state that "ALN are networks created by people learning at any time and any place, using electronic communication tools (...) ALN integrates self-education of an individual with real asynchronous communication with other people. In ALN the student uses a computer and other electronic media and communication technologies to work with far-away learning sources, including in this process the teachers, network administrators and supporting people, but without them being online at the same time". Considering the aspect of distance learning in asynchronous mode we assume lack of online contact between the student and the teacher during a training session. Contact with the teacher is possible during the consulting and the testing sessions.

P. Różewski et al.: Intelligent Open Learning Systems, ISRL 22, pp. 23–50.
springerlink.com

2.2 Comparison of Traditional and Distance Learning

There exist explicit differences between traditional learning and distance learning. They result from the conditioning of the education process and differences in the environment. Different environment, both at the level of education and the level of administration, determines applying a different approach. It is not the goal of the presented analysis to show the superiority of one form of education over the other, since, as was already noticed by many professionals dealing with distance learning, it appears only in the context of a specific application. Individual, dedicated solutions show that a well developed distance learning course can be at least just as effective as an analogous course developed and carried out in traditional learning convention. Courses that fit well in the specifics of distance learning topics are formed of module elements, transferable and open ones. Many examples of such courses can be found in the computer domains, e.g. Cisco Academy. Stress is placed on graphical and audio development. The basis is a well developed scenario based on experience resulting from developing similar courses in traditional form.

A comparison was done by putting together different elements of education, with consideration of the requirements of traditional and distance learning (asynchronous mode).

Table 2.1 Main characteristics of traditional education and distance learning

Main characteristics	Traditional learning	Distance learning (asynchronous mode)
Logical unit	Topics/subject	Learning Object
	The limitations of topics/subjects come from the superior institutions like national education agency.	Based on the domain knowledge structure the learning objects are formed.
Knowledge exchange environment	Direct control	Networked
	The high level of feedback allows to dynamic knowledge exchange environment tuning.	The knowledge exchange process is based on the presence effect.
Availability	The time, place and group of people are all set	Freely place, anytime, for everybody in personalized mode.
Structure of didactic material	Limited context	Modular, open
	The didactical material is limited by the curriculum.	The didactic material organization is based on the module and open structure. The modules can be mixed in almost unlimited way.

Table 2.1 *(continued)*

	Irregular	Regular
Repeatability	The personal teacher approach to teaching process and the random characteristic of learning environment causes the irregular characteristics of educational event.	The earlier prepared learning object linked in the learning scenario allow to achieve constant educational event characteristic. Moreover the approach to content standardization allows to course certification.
Transfer method	Interpersonal techniques, body language, rhetoric	Limited channel
	Direct contact allows to apply many interpersonal communication techniques.	Current technology of communication based on the text and image with multimedia support.
Learning-teaching process pace	Regular rhythm	Explosive nature
	The educational events are organized weekly based on the lesson plan.	Student can entrance to educational event at any time.
Learning-teaching process initiator	Teacher, others students	PULL tools
	The best realization of learning-teaching process is a cooperation between students. Teacher plays the moderator role.	In the Internet system the student is responsible for the all undertake actions
Semantic limitation	Unlimited	Limited
	The teacher usual possesses the knowledge outside the primary domain and he/she is able to answer the student's out of picture questions.	The whole knowledge is included in didactical material. In case, when the student wants to ask the question outside the scope, the system has to call to outside resources (Internet)

In the further part of the chapter an analysis of table 2.1 will be presented in order to describe in detail each of the above-mentioned points. The logical unit point refers to the division of the learning process into elements. Administrative conditions in traditional learning define a clear division of a semester according to the schema: lecture / tutorials / laboratories / seminars / tests. At the level of knowledge learning occurs in deterministic units defined by e.g. the lesson time (usually 45 min.). In distance learning the administrative form is usually similar, while the approach to organizing knowledge is quite different. There are no constraints on time of active learning, since the student can spend any amount of time on a single module. Such assumption remains contradictory with traditional learning, where the entire schedule is strictly defined (this is described by availability in the table). The idea of modules, Learning Objects, allows for organization of knowledge into units, the borders of which are not defined by time, amount of data, but by semantics of the given knowledge. The specific module structure of a distance learning course, as described in the structure of didactic material point, is the indicator of modern information systems. Such systems are flexible, transferable and based on open standards. The opposite is the monolithic structure of traditional learning courses.

The dialogue performed in natural language is the most effective knowledge exchange environment. During the process of knowledge exchange it is important to not only maintain the proper dose of knowledge in the presented content but also to find the optimum tempo of exchange. The limited network bandwidth, being the main communication media in the distance learning system, does not allow obtaining a similar to natural language level of communication. In such case it is important to create the impression of human presence in the given system, so that the student does not feel entirely alienated. Such situation could result in a lot of negative tension, since in such case knowledge may be laden with negative load, what clearly influences the quality of memorizing.

The need for standardization of learning processes at each level of the system's functionality is the ancient problem in the traditional learning environment. We have no influence on the repeatability of the same classes, because there is a random element in them. The teacher is responsible for a given learning process, and like any other human he is sometimes tired or irritated, might forget something or be slightly disorganized. Students co-creating a lesson can themselves be the hindering element in their own learning process. It is highly improbable that we would always provide the same material in the same form and in the same environment. However, these assumptions can be partially realized in distance learning. The repository containing distance learning didactic materials, using the directions of the computer learning support system, can always provide the same portion of knowledge in the same form.

Psychological and pedagogical conditionings are different for traditional and distance learning. It is reflected in the completely different image of the student, appropriate to the educational situation. Transfer in the form of direct contact is substituted by contact over internet. This leads to changes in our behavior. These can be positive changes. Students become more forward and daring, they participate in discussions more actively, take part team work. Also other students become learning initiators, what is very important in case of distance learning. In distance learning the student is responsible for his/her own learning process. All help and support (direct or indirect) leads to improvement of the learning process.

However, there are many more careful voices, indicating the dangers and constraints connected to distance learning (as described for example in [23] and [4]). Creating a good distance learning course requires harmoniously joining such elements as: information technology, the essence of knowledge, psychology, cognitive science. A meaningful conditioning is the transfer method of certain knowledge in a given situation, understood as the richness of techniques, methods and tools used for knowledge transfer. Considering the previously presented information, the difference between networked distance learning and traditional learning seems clear.

Systematic learning is the guarantee of success – it is still one of the up-to-date axioms of education. It is worth considering what being systematic means in the context of differences between traditional and distance learning. Students learning according to the traditional mode attend classes usually once a week. These classes determine the rhythm of learning. The necessity of preparing material for classes and the possibility of being tested from already completed material, force the student to work systematically. During regularly held classes, it might happen that the student will miss a class. Excessive amount of work on a given day might cause the student to be tired and result in lack of focus and attention. These disadvantages do not occur in distance learning, where students work with didactic material when they are prepared to give it enough attention. This freedom leads to extremes, when students once a month or less try to catch up with the missed material by looking through it all. Both the traditional and the distance approach require students to work systematically. However, distance learning, due to its specificity (individual work over internet), sets higher requirements for the student. Therefore different techniques are applied to help the student, e.g. by organizing intermediate activities and tasks within the semester or by temporarily limiting access to individual parts of the material, like in a time window (e.g. certain material is available only for two weeks).

Human adaptive capabilities allow for dynamic expansion of the context of expressions. Depending on the specific situation, the teacher can adapt the prepared material to the listening students. That often requires expanding the material with additional information that is especially interesting and important for the students in the context of their interests or work. Didactic material prepared for distance learning is a formalized representation of a certain area of knowledge and as such is limited and closed. As a way out it is possible to use internet resources. However, this leads to many inconveniences connected to lack of influence on the credibility of presented information and the possibility of losing integrity of the transferred material.

Differences between traditional and distance learning can be observed at different levels. Especially interesting is the identification of characteristics of students learning the same subject through traditional and distance learning. Studies documented in [2] show differences occurring during observation of students studying a course dedicated to basics of the C++ language. Students studying over distance are usually older. They treat distance learning as vocational education expansion. Usually they already have parental duties. They can be characterized by lesser knowledge of computers, but on the other hand by greater awareness of the goals that they want to achieve. Considering the American specificity of the presented studies, it is easy to notice sociological differences that arise between traditional and distance students. Students using distance learning, as was shown in [12] and [13], obtain new abilities in working with a computer, learn to communicate and to cooperate with other people over internet, have the ability to obtain education they did not have due to their family, financial or vocational situation.

As was proven by the analysis presented in table 2.1, distance learning and traditional learning possess both positive and negative characteristics. The task is to develop a solution that avoids disadvantages of traditional learning, including problems caused by the defined rhythm and location of meetings, while at the same time maintaining its advantages that result from direct contact between the teacher and the student. To solve the considered task, the use of modern computer technologies should be maximized. Finding an appropriate approach to such a widely stated task requires detailed analysis of the learning- teaching process from the point of view of:

- subjects, goals and means of education;
- participants of the processes and roles played by them;
- content and order of operations consisting for the learning process.

2.3 New Environment for Distance Learning

2.3.1 The Role of Internet in Education

Before the new environment for distance learning will be discussed, the following question must be answered: which is the role of internet in education: to replace or to assist the teacher? We try to answer this question through performing an analysis of properties and features of both: the human teacher and the interned powered computer. As a starting point for our consideration we used a well known proverb:

Knowledge consists of information
like a building consists of bricks.
But not every couple of bricks forms a building
and not every collection of information forms wisdom.

The analogy between two processes: construction of a building and forming of wisdom in students brains is very useful for the discussion of the roles of computers and the human teachers in education (fig. 2.1).

As an information- carrying medium, Internet is very useful for the main objective of teaching when we talk about equipping the students with certain knowledge. It is very similar to transporting the bricks and other construction materials for the building (see leftmost part of fig. 2.1.). After the necessary amount of construction materials is collected, architects and construction engineers, who can and know how to form materials into the building structure (see central part of the fig. 2.1.), must come. Educational analogy for this process

Fig. 2.1 The analogies between construction of the building and education process

is the professor establishing and explaining dependences and relations between separate pieces of information in the frame of a general knowledge system. The problem solved here is connected with the fact that very often students can have a big collection of information but cannot alone perform the internalization process, which is in fact absolutely necessary for transforming the raw information into firm and reliable knowledge in the students' minds.

The role of the human teacher is very important in the inspection process – both in construction procedures and in educational processes. Every education must be connected to a quality assurance process, which must be involved also into the graduation scheme. Here the participation of the human teacher is not simply necessary, but rather indispensable. The very important area of all educational activities, restricted for human teachers only, is covered by many types of exams and testing procedures measuring of knowledge and qualifications (see rightmost part of the fig. 2.1.).

On the basis of the previous consideration we can give a more detailed description of the human role in the computer-powered distance learning-teaching process. The role of the education process is not only to give students a collection of more or less useful information. One of the goals of education is also rearing the students in accordance with some approved ideals so that they acquire and respect the universal academic values revolving around the transcendental triad: truth, goodness and beauty.

Each of these enriches a different aspect of human nature. The truth improves the intellect while goodness and beauty shape the will and emotions. The problem of values is of primary importance in present-day education as they are closely connected to the objectives and goals of education. Regardless of the actual level, the main task of each school is ensuring learners' development in terms of values, knowledge and abilities.

Even most dedicated advocates of the internet do not dare to claim that this tool can bring anything significant in the sphere of values. It is now a questionable point how much one can expect from new tele-information tools in the field of knowledge transfer and development of students' abilities. The conclusion is that only a certain portion of the teaching job can be left to the internet. It would be worthwhile to reflect on the boundary line in the education process, between the part that can be assisted by well-programmed machines and that part where an experienced and qualified teacher becomes necessary.

It seems that in modern schools the traditional relations between the master and his disciples are quite lost. This relation allowed pursuing an admirable goal: education leading to wisdom. Present-day education is becoming more and more concentrated on pragmatism, efficiency, and performance, winning a high position on the labor market. The education process is not oriented at wisdom, imagination, humanism and values, it does not promote disinterested knowledge. It is in fact, nearly the production of specialists, the same as manufacture of industrial products. N. Postman remarked that the basic goal of education in technocratic societies is constant improvement of efficiency and attractiveness of the teaching process and the only justification for these actions is the cliché that "education helps students to find good jobs".

This is well illustrated by a remark made by T. S. Elliot, an English poet and laureate of the Nobel prize, who asked[1]:

Where is the wisdom we have lost in knowledge?
Where is the knowledge we have lost in information?

This sentence allows for formulating the ordering sequence:

information < knowledge < wisdom

This notation means that knowledge means something more than information, while wisdom is still more than knowledge. Information is the necessary foundation for knowledge while wisdom "does not mean the same as knowledge though it assumes that one has it at least in the extent ensuring correct behavior in the given conditions and circumstances in which we happen to live".

The direct result of teaching is the growth in the level of knowledge. Information changes into knowledge in the process of internalization, when proper structures are developed in the human mind. In that process new data is incorporated in the existing resources, which are thus extended, though not by way of simple additive incrementing of the existing knowledge, but also through

---

[1] Source: http://en.wikiquote.org/wiki/T._S._Eliot

independent formation of new semantic values based on mutual relationships, correlation and synergy of elements. The major point in the process of internalization, when information is changing into knowledge, is that knowledge is a system made of bits of information. The general theory of system has it that a system will always be something more than a simple sum of all its elements.

In fact the problem is much more complicated. As you can see in fig. 2.2. dependences between at least four things must be described: education, information, knowledge and wisdom. In fact, it ought to be the object of scientific research for many years and for many researchers.

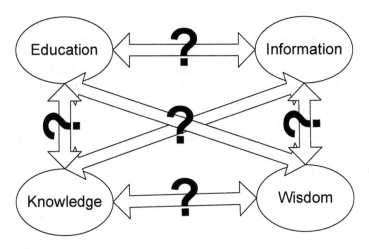

Fig. 2.2 Some important dependences, which must be defined when talking about the distribution of roles of man and machine in the new e-learning process

Continuing the discussion about the internalization as the necessary condition for development of knowledge, we emphasize the fact that this process always ensures cohesion of thus acquired knowledge. As a result, the learner will try to eliminate contradictory bits of information, bits of outdated information will be replaced with the current ones and detailed information will be replaced with more general one. We have to emphasize that this process should be treated subjectively, it is an activity of the learner and the presence of a teacher may facilitate and speed it up. This view is confirmed by C. Freinet, who remarked that contrary to what people sometimes think, knowledge is acquired by experience, not by learning the rules. Thus a self-taught person can have considerable achievements though the experience of nearly all school systems all over the world clearly shows a positive role of qualified teachers, since the process of data acquisition is faster, more effective and the information is better remembered.

In the humanities, information is regarded as a message or contents passed on in the process of communication between people. A more formalized definition of this term is adapted to the requirements of information technology, yet we will not refer to the mathematical theory of information even though it might prove useful for development and optimization of the means of interpersonal communication, especially for creating systems in which data processing and storing is done by machines.

"Knowledge" is a more complex term. The definition provided by the dictionary is as follows: "meanings and concepts imprinted in human mind as a result of gathering experience and learning". Two categories of knowledge can be distinguished: experience- based practical knowledge, which provides information on how to change the reality, and theoretical knowledge providing information about this reality.

The last component is the most difficult to define, it is not easy to find an adequate and satisfying definition of wisdom. This opinion is borne out by the fact that the definition is often sought in literature. All available materials lead to two major sources: dissertations of ancient philosophers and religious writing. Analyses of such sources help us define the most important attributes of wisdom:

- striving for truth
- ability to make independent and right choices
- ability to differentiate between important and trifle things (valuation, building a hierarchy of things)
- life-long experience
- "ethical spirituality", honesty, responsibility, upright character, high ideals
- desire to ensure good and long life
- consistency of thinking and acting to solve new, not typical and unique problems

When the learning-teaching process is perceived in a wider perspective, as assisting people in their personal development and formation, to help them grow in wisdom and live by the values, then we can ask a question about the role of internet in achieving this goal. At the same time we are faced with the question which part of the teaching process can be supported by computer and which has to be left in the hands of teachers. This leads to a more concrete question: the internet in education or education in the internet?

Analyzing the stages of the quest for wisdom, we come to the conclusion that information can be transferred by way of contacts between man and machine. Moreover, the didactic material and learning environment development according to cognitive requirements allows to transfer the knowledge as well. Therefore, while searching for the place of internet in education, we should focus on the aspects relating to information and knowledge transfer.

2.3.2 New Role of the Teacher

In the conditions of distance learning, conducted in asynchronous mode, student obtains knowledge without direct contact with the teacher, as well as with limited contact with other students. Distance learning in asynchronous mode assumes

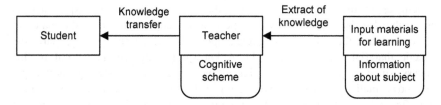

Fig. 2.3 Organization of the cognitive process in traditional learning (based on [11])

division in time and space in the contact between the teacher and the student [22]. This division implies necessity to change the traditional methodology and adapt it to the new educational situation. One of possible solutions to this problem was presented in [11]. The method proposed by the authors of that paper assumes co-existence of three components: teacher, student and didactic material in the frames of the cognitive process of the student. At the primary stage of the preparation process the teacher selects knowledge form the available material and uses possessed knowledge to create the domain's cognitive model. The cognitive model being created depends on the defined learning goals and student's basic knowledge. The teacher creates a bridge, in the form of didactic material, between the student and the knowledge connected to the domain. In the traditional learning model, depicted in fig. 2.3., the teacher is the organizer, initiator, care-taker, adviser, authority, whose goal is to lead the student to the learning goal, taking into account his/her motivation and other pedagogical constraints. The teacher is responsible for creating and developing the cognitive model, according to which knowledge is transferred to the student. That means that quality of the education process strongly depends on the teacher's ability to build the cognitive model.

The basic indicator of the asynchronous mode is the occurrence of the learning process without direct contact with the teacher. Additionally – the learning process itself is based on a free time regime. This means that the basic method of learning is self-education. New computer technologies and the state of development of telecommunication networks give possibility for direct contact between the teacher and the student regardless of the distance existing between them. However, applying these techniques is connected to high costs related to the network infrastructure. Another approach relies on transferring the conventional textbook to the telecommunication space of distance learning without interfering with the cognitive process. In such case ensuring proper effectiveness of education requires building and online dialogue. Otherwise the learning process is ineffective.

In distance learning the teacher does not perform the same function as in traditional education (he does not directly transfer knowledge). In such case it is reasonable to form research questions: what is the new role of the teacher, how should the didactic materials be changed (adapted)?

Distance learning changes the role of the teacher and didactic material in the learning process. Didactic materials acquire the characteristics of a knowledge base, what results in the teacher on one hand becoming an expert in a given domain and on the other hand – a consultant and an examiner. The teacher participates in the process of creating didactic material as an expert and is responsible for the content-related scope of knowledge included in the material, and, considering the defined pedagogical approach and learning goal, prepares the didactic material (fig. 2.4., Step 1). A knowledge engineer is helpful in this process, since he possesses skills that allow forming and modeling knowledge, and knows methodologies of constructing distance learning courses. In the next step (fig. 2.4., Step 2) the teacher functions as a consultant and examiner of student's knowledge. Didactic material is divided in two parts: knowledge of the domain of the learning subject and cognitive model connected to it. The student participates in the learning process by independently working on the didactic material in the frames of the distance learning system. The teacher, whose responsibilities now have a different character, can take care of a bigger number of students. The approach assuming a new role of the teacher blends in with the theory of constructivism. The form of problem-based learning places the teacher in the role of just a scientific advisor.

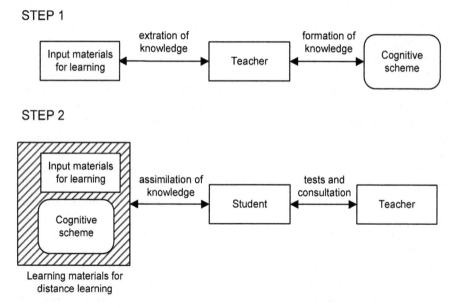

Fig. 2.4 Model of the education process considered from the point of view of learning over distance (based on [11])

The most difficult challenge connected to introducing the technology of distance learning is the need to spend a lot of time both on training teachers and students in computer techniques. Effective use of distance learning over internet requires possessing a proper amount of knowledge and skills. As was shown by studies regarding the development of distance learning in South Africa, described in [1], lack of technological knowledge may be the cause for a significant decrease in effectiveness of learning. On one hand, in countries that have problems with educating their society (e.g. due to the distance or lack of staff) distance learning is a great tool increasing the level of education, but on the other hand, we encounter there a barrier in the form of lack of technological knowledge.

Another thing worth analyzing is the character of the need for specific specialists taking part in distance learning in the given institution during consecutive phases of realizing the project. Through analyzing individual phases of realization, implementation and exploitation of the distance learning environment we can state, according to [3], what skills are stressed in each phase. The initial activities are always connected to the work of network, telecommunication and software specialists. They create a working system, which is then filled with materials created as a result of work of specialists of methodology, design and creation of didactic materials. The exploitation phase, the longest of the phases, requires activity oriented on marketing and promotion, budget, planning, correcting the program and materials, graphic and software works, administration. The teacher is in his efforts supported by the created systems of distance learning, which, basing on mechanisms of artificial intelligence, perform more and more responsibilities, tasks and also more advanced functions (e.g. recognizing the student's learning style, intelligent administration).

2.3.3 Collaboration

One of the elements increasing effectiveness of distance learning is collaboration between the student and the computer system. The environment of distance learning provides a proper collaboration platform only when it ensures the following, described by [5], features:

-   the student has to know how he/she will benefit from collaboration with other students or with the system;
-   the environment must be comfortable for the student and arise trust;
-   the student should trust the teacher, the instructor and the system;
-   the student should feel that he/she is immersed in a rich, involving environment that supports contact with other students and the teacher.

Showing the student the values and methods of improving collaboration encourages him/her to increase own initiative in this area. This can be achieved through showing advantages that other participants achieve together with the interested student. A helpful thing might be conducting an action of information distribution where advantages of collaboration are clearly visible. The system should be sensitive to noticing every, even very small, display of participation in the prepared program of collaboration. Creating effective collaboration may encounter problems connected to the applied technology. Transmission loss, delayed packages arriving in a wrong order, problems with identification and authorization of the content transfer may cause resistance regarding participation in collaboration during the learning process. Providing a proper platform increases trust in collaboration. It is equally important to create trust at the level of collaboration between certain people, in this case: students and the teacher. Excellence of identification, availability and transparency of the presented information, together with flexibility regarding its edition and preparation, allow for achieving this goal.

2.3.4 New Requirements Regarding the Structure of Didactic Material

We can state that didactic material has always been the most important mean of learning. The importance of didactic material increases significantly in distance learning conditions. The reason for this is the need to reflect not only the content, but also the cognitive model.

Obtaining low cost of education, possibility of contact over distance, and no decrease in the level of effectiveness of learning requires re-organizing the cognitive process. The cognitive model that is created in the mind of the teacher should be presented in a formal way and should become a real part of the didactic material. The carrier of the model in traditional learning is the teacher. In case of distance learning there is a need to isolate the cognitive model, formalize it and implement in the structures of didactic material. The most difficult stage is the formalization of the cognitive model. Modern computer technologies can be of help here, allowing for visual manipulation of structures, facilitating (through metadata) building relations and relationships.

Distance learning does not cast aside traditional sources of knowledge, such as: monographs, textbooks, articles. The content passed on in the distance learning education process is the same as in the analogical course, with similar education goal, prepared for traditional learning. However, didactic material faces new, special requirements that arise from evolution of traditional knowledge sources.

Didactic material dedicated to distance learning, also known as courseware, allows students to learn over internet, independently from the teacher. The basic structure element of courseware is a conceptual schema, the goal of which is to present concepts and relations between them [11]. This enables organization of knowledge at the level of concepts. The conceptual schema is a structure the goal of which is to integrate content with appropriate cognitive model.

The courseware material has network structure (fig. 2.5.). Nodes of the network are the basic learning concepts. All concepts are placed in one network structure. Characteristics of each concept, defining its context, are placed in a dictionary. When context of a concept contains reference to another concept, the dictionary has a hypertext structure. In the course each concept is connected to a computer metaphor, which is considered a multimedia representation of the concept. Very often each concept has a reference to its original definition.

The schema of didactic materials' structure presented in figure 2.5. reflects the content of the conceptual model. The conceptual model considered in the context

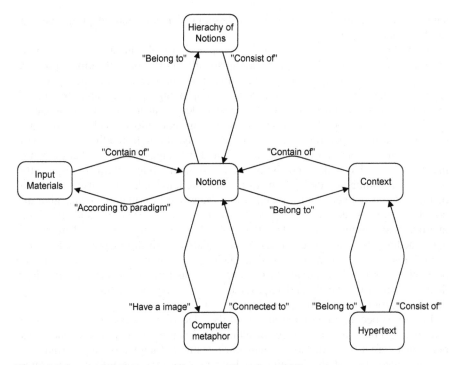

Fig. 2.5 Schema of didactic material structure (based on [11])

of a specific domain is reflected through ontology [7]. Each concept used in courseware belongs to a structure of concepts that does not exceed the chosen paradigm. In case of knowledge that can be based on a taxonomic model, its reflection has a hierarchical structure [8]. Each concept is defined by a computer metaphor. The computer metaphor provides visual representation of the concept on the basis of multimedia capabilities of modern computer systems. The way of using technology consists for the created illusion, the goal of which is to pass specified educational content. The current technological level allows building advanced courses using 3D graphics and other methods of reflecting reality and creating virtual learning environments.

2.3.5 Influence of Network Environment on the Concept of Distance Learning

Network environment requires new organization of student's work. It is caused by the change in the work concept, where the learning and teachings sides are divided by space and contact is based on telecommunication or computer means, through a properly organized knowledge exchange space. In this case, accents are placed on creating the impression of direct contact with the learning system and other students. That requires applying new technological means and adapting existing teaching methodologies.

As we try to show in the following chapters, virtual education can be more personalized, but it not happen automatically [14]. In education, process presentation and deployment of teaching materials through information networks, especially internet, is the natural way of improving quantity and quality of learning. Before the internet, the only ways to learn outside the classroom were enrolling in correspondence courses, watching educational videos, software designed around specific content, or self-study through reading books and journals. Now it can be done easier, faster and in a bigger scale – what does not mean however that it does not involve any problems. Internet can be especially useful in spreading detailed information and basic knowledge, because it is a tool of information delivery and dissemination. New tele-information media give all possible students access to extensive sources of information and facilitate interpersonal communication [15]. However, they reject the traditional model of mass media, where the receiver had to be satisfied with things presented by the central distributor. Interactive media, such as the internet, give the user the possibility to control the stream of information, thus forcing the suppliers to adapt their offer to users' needs and requirements. It is very promising but also very demanding for organizers of virtual learning a teaching process.

Internet is becoming an integral part of the education process, playing a double role of the subject and the object of educational activities [16]. The connections between internet and education seem fairly natural. Internet is an information-carrying medium and the learning-teaching process involves information. Thus using internet as a new teaching tool may improve the efficiency of information transfer and accessibility, and in consequence it should positively impact all areas connected with teaching and learning.

2.4 Simplified Cybernetic Model of the Learning and Teaching Process

The proper combination of e-learning system and computer-based teaching materials and personal contact between students and teachers (masters) can be explained using a cybernetic model of the learning-teaching process, described for the first time many years ago in [21], next developed in [18,17,20] and in final version described in [9,10,24]. The model is formed in terms of cybernetics, because all processes which can be taken into account during scientific analysis of learning and teaching processes are of information nature, and cybernetics offers the best tools (and best language) for describing and for analyzing every information phenomena. The next strong advantage of cybernetics, related especially to the problems considered in this paper, is connected to the facts, that cybernetics enable unifying the form of description of similar information processes performed by different subjects (e.g. human teachers and computers).

Studies show that using the computers and internet for educational purposes will not eliminate teachers, but will lead to a change in the teacher's role: instead of being the supplier of knowledge the teacher will become an organizer of the education process and an experienced guide helping learners to structure the knowledge they gained from various sources, including internet. When the expensive model of education: "Teacher <> Student" gives way to the cheaper model "Computer-Student", the teachers will no longer have to deliver all the information, which is an arduous and barely creative task. Teachers' activity will then focus on development of knowledge and showing its practical applications. With regard to taxonomic levels and categories, we can say that internet has a positive function at the lowest levels of taxonomy, i.e. at the level of information gathering, while training students to use this information in new or problematic situations will still be the domain of teachers.

This task is extremely important in the light of doubts expressed by those who are afraid that multi-media and electronic carriers of information will produce a society capable only of superficial judgments, as surfing in the cyberspace and in virtual reality does not require any intellectual effort or analytic thinking, neither does it force a man to be patient or to concentrate. Furthermore, a man may become a slave of his own tools.

The model of internet-assisted education, presented in this study, seems most interesting from the point of view of cognitive approach and economic considerations. This concept, presented very briefly, is a part of distance learning - a new and fashionable approach. Introduction of internet-assisted education has to be done very carefully and the teachers have to be trained so as to benefit from these new facilities. The training should include searching and selection of valuable information in the internet, as well as the ability to develop their own teaching materials and presenting them on the servers. These goals, postulating that each teacher should have the basic skills in operation of information tools, are set forth in the document "Objectives", prepared as a part of the reform of the education system in Poland.

To analyze the proposed cybernetic model of the learning-teaching process we started our considerations from a basic model (Figure 2.6).On this schema it is visible that in the traditional framework of the human-based teaching and learning process the structure and relations form a well known cybernetic structure: a simple feedback loop.

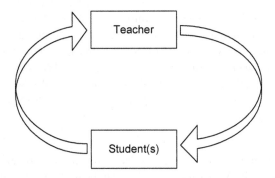

Fig. 2.6 Traditional human-based model of learning and teaching based on a single feedback loop schema

There are two streams of information in this system. The first on passes from the teacher to the students: facts, knowledge, information, presentations, interpretations, explanations, questions, tests, etc. In the model for all this information we try to use only one symbol I, standing for Instructions, Information, Inspiration, Inspection etc. Taking into account the dynamic character of the teaching process we use the function $I(t)$, where, for convenience, the discrete time scale $(t = 1, 2, \ldots\ldots , E)$ will be applied. Points on the timescale can be connected to the moments of consecutive lectures, and the end point E based on the constant time diapason – for example to an officially established duration of the semester.

From the student to the teacher the following can be passed (taken by the teacher): answers, test results, activity evaluations, asked questions, general evaluations, abilities for problem solving, etc. For all these activities and evaluations we use in our model one symbol K (from Knowledge), which is a function of time $K(t)$. In a good education system $K(t)$ permanently increases.

For modeling (and for computer simulation of the model) some quantitative sense for $I(t)$ and $K(t)$ must be given. Let us assume that some measure of the amount of information given by the teacher to the student can be found, and let the value of this information volume be denoted by $I(t)$. This is a very general view, in details the information stream going from the teacher to the students must be described using many quantitative and qualitative factors – but for the purpose of this study the symbol $I(t)$ is used as the first order approximation of all things under considerations.

For analogy, let $K(t)$ be considered as the measure of student's knowledge. The value of $K(E)$ (value of knowledge collected by the students at the end of the education process) may be then shown as the criterion for evaluating quality of the entire learning-teaching process.

The cybernetic model must also include a mathematical descriptions converting $I(t)$ to $K(t)$. From the point of view of the description of the process carried out by the students during learning, it becomes clear that a relation must be established between K and I. In real life this relation is very complicated, but considering the simplest situation we can assume that the volume of knowledge K obtained by the student is proportional to the amount of information passed from the teacher to the student. Let S (from student) be the symbol for the proportion coefficient and let us assume that the process of internalization of the knowledge in the student's brain has its own dynamics, what results in a one-step delay between the moment when some piece of information $I(t)$ is given by the teacher to the student, and the moment when it can be transformed to knowledge $K(t+1)$ in the student's brain. Such assumptions can be written in the form of an equation:

$$K(t+1) = S\, I(t) \tag{1}$$

According to the above equation, the smartest students will be characterized by the highest value of the S coefficient, as they can better transform the information $I(t)$ obtained from the teacher into their private knowledge $K(t+1)$.

In a similar way a simple equation can be used to describe all relations and dependencies (very complicated in reality) between diagnostic and didactic information, obtained by the teacher by means of testing procedures and exams, representing the measure of actual knowledge of the students $K(t)$, and the volume of information $I(t)$ passed to the students at a step of the teaching process. The equation can be written in the form of

$$I(t) = T\,K(t)$$
(2)

where coefficient T represents the teacher's diligence, because a bigger value of T means, that the teacher gives more information $I(t)$ to the student in the same situation, characterized by the actual value of the student's knowledge $K(t)$. According to the model proposed above, teacher's adaptability (in terms of adjusting the information given to the students to the actual level of knowledge of these students) can be very quick in comparison to the process of knowledge acquisition in students' brains.

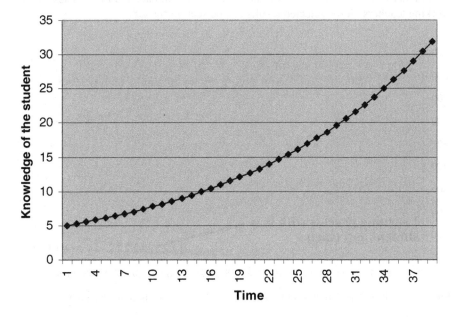

Fig. 2.7 Result of simulation performed (based on the proposed model)

Putting together equations (1) and (2) we obtain a formula that can be used for describing dynamics of the knowledge accumulation process during learning and teaching:

$$K(t+1) = S\,T\,K(t)$$
(3)

Using this formula we can calculate (starting from an arbitrary initial value $K(0)$ a whole chain of values $K(1)$, $K(2)$, ..., $K(E)$, describing, in quantitative form, the dynamics of the main result of every learning-teaching process: increase in students' knowledge (see fig. 2.9). The general form of the curve plotted in fig. 2.9. is acceptable from the point of view of teacher's practice: the effectiveness of learning increases with the growth of the total knowledge of students. Some details (and difficulties) connected to this fact will be discussed later, but the general form of the model's behavior is acceptable.

2.4.1 Interesting Results of Simulations Performed Using the Proposed Model

The model under consideration is the most simplified one, but on the base of this model many interesting forms of the learning processes can be shown, by means of computer simulation for different parameters included in the model. As mentioned above, the learning process can be simulated for more or less hard working students with higher or lower initial level of knowledge and potential, and more or less hard working teachers. Some selected results can be seen in fig. 2.8.

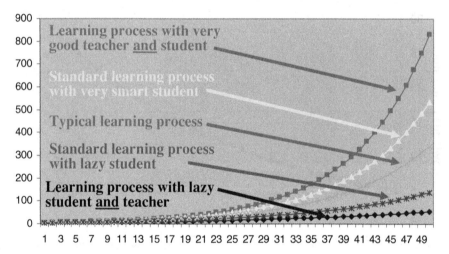

Fig. 2.8 Results of simulation for different values of parameters used in the model

Results of the simulation are logical and self-explanatory: the best results of learning and teaching one can obtain when the teacher is very hard working and the student is very smart, while a similar process performed by a lazy student and a lazy teacher leads to almost zero-growth of knowledge. Sometimes during a very bad education process (small T and S values) the student can in fact lose his (or her) initial knowledge, as can be seen through the simulation results presented in fig. 2.9.

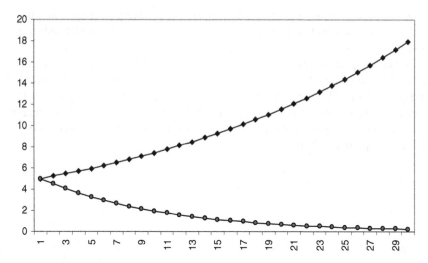

Fig. 2.9 Simulation of a very bad learning process, leading to the loss of student knowledge

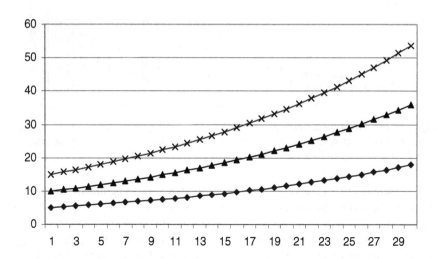

Fig. 2.10 Simulation of the learning process which starts from different initial levels of student's knowledge

The model can also be used for analyzing relations between the initial knowledge of students and the obtained results of education. In fig. 2.10. it was shown how important initial knowledge can be for the effectiveness of learning and teaching. Using the model, we can obtain many other results of simulations, but in fact all these results are trivial (anyone can predict such relations using common sense, without any simulation). Nevertheless, every simulation result can be found as logical and interpretable, what proves that the entire model, although a simplified one, is acceptable in general.

A very important result, obtained using the model under consideration, is connected with the observation of general nature, common for all simulated processes. As can be seen, all the learning curves are of exponential shape. It is because of the simplifications included in the models (linear form of dependences) and because of the general form of the one-loop structure of all models, based on the scheme presented in fig. 2.6.

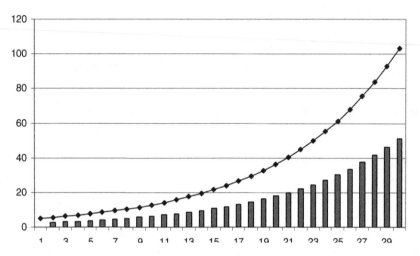

Fig. 2.11 The plot of the learning process performed according to the exponential law (blue line), the consecutive pieces of information (brown blocks), represent the information which must be accepted by student till next lecture

Exponential curves are very useful in theory, but have some inconveniences in practice. According to the exponential rule, the amount of new information a student should accept from one step to another permanently increases (fig. 2.9.). This means that the student's effort must permanently increase during the learning process – and this fact can be a source of problems. Not all students can be so efficient in their work. Also, personal problems of the students, e.g. connected to some illnesses, accidents, money worries, family problems, happy or unhappy love, and many other objective and subjective difficulties can lead to a learning crisis, as shown in fig. 2.12.

When this crisis point appears, the only way to efficiently help the student is to adapt the actual acceleration of the teaching process to the actual student's mental and psychological abilities (fig. 2.13). Except that it is impossible according to the exponential rule (fig. 2.9.) describing the teaching process in the classical model (fig. 2.6.).

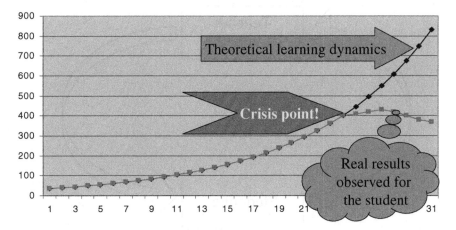

Fig. 2.12 Crisis resulting from a too quick learning process performed in a single loop educational model

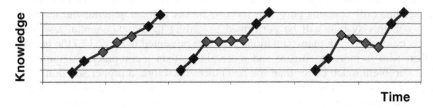

Fig. 2.13 Adaptation of learning process to the current capabilities of the student

2.4.2 Cybernetic Model of a System Formed by Teachers, Students, and Computers

A solution to the problem pointed out at the end of the previous chapter is possible once a proper place for the computer in the IT-based education system is found. The proper place means that it is not a good solution to simply replace the teacher by the computer in a classical one-loop learning scheme (fig. 2.14.). Many (or even most) of the so-called "modern e-learning systems" or "virtual universities" are designed and implemented in the form presented below, and exactly this fact is one of the most important reasons for well known weaknesses of these typical e-learning systems.

The type of system as presented in fig. 2.14. is very popular, since the education process performed according to the presented structure may be very cheap and (in simple cases) very efficient. However, in more complicated situations, especially in higher education, this system seems to be unsuitable, although it still remains realizable from the technical point of view. Let us consider the weaknesses of this popular structure.

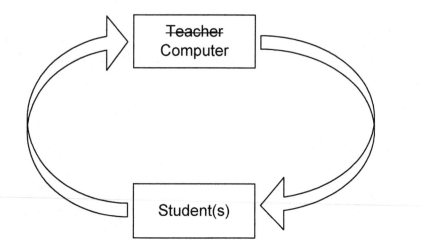

Fig. 2.14 Simple replacing of the teacher by a computer is not a good solution

The first and most important weakness is connected to the fact that this "new-old" schema still contains only one loop. It means that, in general, the learning process in this system will be performed according to the exponential law and, what is much worse, in many situations (e.g. like presented in fig. 2.12.) the human teacher, who is more intelligent and more elastic than a computer program, can solve a problem much better than the computer. Therefore, a proper solution for a computer-aided teaching and learning system must include both a human teacher and IT machines, and must divide the tasks between these two components.

In order to determine the optimal solution, let us first try to add a computer to the traditional structure as presented in fig. 2.6. For the purpose of this study, we are not taking into account trivial (and thus very popular) applications of computers in the education process, in which the computer is used only as a tool for preparing and showing presentations, thus only providing better visualization than traditional lectures, based only on blackboard and chalk (fig. 2.15a.). We also eliminate from our considerations all systems in which computers (and other IT elements connected to them, like internet) are used only as additional sources of information for the students (fig. 2.15b.) or as a tool for testing students' knowledge, when computers are used as test-serving and answer-collecting machines, sometimes powered with statistical processing of the data and with many helpful utilities, used for easier and more efficient evaluation (fig. 2.15c.).

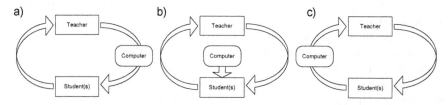

Fig. 2.15 Three popular places, in which a computer can be attached to the education system without changing the general structure of the system

Computers can be very useful in each of the places depicted in figure 2.15., but neither of the presented structures changes the general form of the one-loop feedback system, and cannot serve as the source of new quality in learning processes. The really important step towards new quality in an electronically powered learning-teaching process can be performed when we organize a system in which the computer helps forming two additional feedback loops. Such two loops are formed inside the traditional learning structure and consist of the first (fig. 2.16.) loop formed by interactions between the student and the computer (a good interactions between the student and the computer must be arranged) and the second (upper) loop formed by interactions between the teacher and the computer.

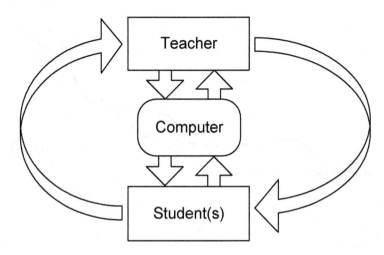

Fig. 2.16 The best (and recommended here) learning and teaching system, consisting of three closed loops formed by proper connections between the teacher, students, and computers

The role of the first inner loop (between the computer and students), presented in fig. 2.16., is well known, as there are many valuable papers considering methods of better cooperation between man and computer (see for example [19]). Therefore one might say that from the scientific point of view everything is clear: good interaction is crucial for good application of computer-based learning. A computer cannot be simply an electronic book for reading, a program intended as a good education tool must be interactive. Computers ought to force and push students to use obtained knowledge. Every part of information served to the student must be connected with test, quizzes, tasks and questions for students. One other very important part of every good education program implemented in the teaching computer must be a knowledge evaluating mechanism, with automatic adaptation to the student's personal preferences and abilities. Automatic testing must be a friendly, interesting and nice process, nevertheless, the student cannot proceed to the next step of education if automatic inspection discovers that his (or her) knowledge is still below the desired level in quantity and quality. Using the computer in such a way as an additional source of information (and incoming knowledge), the education process can be sometimes sped up, and in other moments slow down, according to the student's changing abilities and possibilities. As a result, we obtain a system in which every student can come his own way to the desired level of competences, and this system is completely secure against educational crisis (fig. 2.17.).

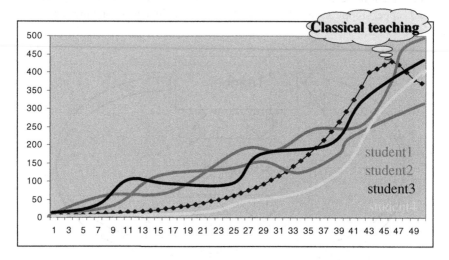

Fig. 2.17 Individual traces of education in a multi-loop e-learning system versus the traditional model

A more complicated situation one can find inside the second inner loop, connecting (in frames of feedback contour) the human teacher and the computer.

2.5 Conclusions

The conditioning of distance learning does not allow for direct transfer of didactic materials created for traditional learning to the space of distance learning. The task of building a distance learning environment focuses on solving new problems, the source of which is the concentration of the learning-teaching process around the issues of cognitive conditioning of a student. Knowledge of the mind is the main stimulus defining the new structure of didactic materials and having influence on the process of distance learning.

The presented analysis shows that creating a detailed model of processes consisting for an educational organization is difficult. The main reason for that is the individual character of separate educational organizations.

References

1. Arif, A.A.: Learning from the Web: Are Students Ready or Not? Educational Technology & Society 4(4), 32–38 (2001)
2. Dutton, J., Dutton, M., Perry, J.: How do Online Students Differ from Lecture Students? Journal of Asynchronous Learning Networks 6(1) (2002)
3. Hill, M.N.: Staffing A Distance Learning Team: Whom Do You Really Need? Online Journal of Distance Learning Administration 1(1) (1998)
4. Hokanson, B., Hooper, S.: Computers as cognitive media: Defining the potential of computers in education. Computers in Human Behavior 16, 537–552 (2000)
5. Hughes, S.C., Wickersham, L., Ryan-Jones, D.L., Smith, S.A.: Overcoming Social and Psychological Barriers to Effective On-line Collaboration. Educational Technology & Society 1(5) (2002)
6. Juszczyk, S.: Distance learning education: concepts, rules and processes. In: Edukacja na odległość: kodyfikacja pojęć, reguł i procesów. Adam Marszałek Publishing House, Poland (2002)
7. Kusztina, E., Zaikin, O., Różewski, P.: On the knowledge repository design and management In E-Learning. In: Lu, J., Da Ruan, Zhang, G. (eds.) E-Service Intelligence: Methodologies, Technologies and Applications. SCI, vol. 37, pp. 497–517. Springer, Heidelberg (2007)
8. Kushtina, E., Różewski, P., Zaikin, O.: Extended ontological model for distance learning purpose. In: Reimer, U., Karagiannis, D. (eds.) PAKM 2006. LNCS (LNAI), vol. 4333, pp. 155–165. Springer, Heidelberg (2006)
9. Kusztina, E., Zaikin, O., Rozewski, P., Tadeusiewicz, R.: Conceptual Model of Theoretical Knowledge Representation for Distance Learning. In: Dijkman, H., Veugelers, M. (eds.) Beyond the Network – Innovative IT-Services, vol. 2003, pp. 239–243. EUNIS, Universiteit van Amsterdam (2003)
10. Kusztina, E., Rozewski, P., Zaikin, O., Tadeusiewicz, R.: Distance Learning Organization based on General Knowledge Model. In: Ribeiro, L.M., dos Santos, J.M. (eds.) The Changing Universities: The Challenge of New Technologies. New Technologies for Teaching and Learning, EUNIS 2002, FEUP, Lisbon, pp. 401–406 (2002)
11. Kusztina, E., Zaikin, O., Enlund, N.: A knowledge base approach to courseware design for distance learning. In: Knop, J., Schirmbacher, P., Mahnic, V. (eds.) The Changing Universities - The Role of Technology, The 7th International Conference of European University Information Systems. LNI, Berlin, Germany, vol. 13, pp. 499–505 (2001)
12. Spiceland, J.D., Hawkins, C.P.: The Impact on Learning of an Asynchronous Active Learning Course Format. Journal of Asynchronous Learning Networks 6(1) (2002)
13. Tadeusiewicz, R.: Internet Community (Die Internetgemeinschaft in German). Shaker Verlag GmbH, Aachen (2005)
14. Tadeusiewicz, R.: Virtual Teaching on the Basis of Experiments in Computer-Assisted Instruction at the University of Mining and Metallurgy of Cracow. In: Higher Education in Europe, UNESCO CEPES, vol. XXVI(4), pp. 553–566 (2002)
15. Tadeusiewicz, R.: Technical Education and the Demands of the Information Community – The Opening paper for the 23rd Polish Engineers Congress, Foundry Survey, vol. 2, pp. 45–50 (2002)

16. Tadeusiewicz, R.: Selected Threats Resulting from Use of Internet in Teaching. In: Conference Proceedings of the 9th Polish-Wide Scientific Symposium Computer Techniques in Education Transfer, Kraków, pp. 73–91 (1999) (in Polish)
17. Tadeusiewicz, R.: Computer Methods in Education as One of Characteristics of Information Society. In: Woźniak, J., Miller, R.C. (eds.) Research Libraries: Cooperation in Automation, vol. 3, pp. 11–22. SBP Publisher, Warszawa (1999) (in Polish)
18. Tadeusiewicz, R.: Computer Aided Teaching. Teachers Preparation in Computer Science, Problems of Teachers' Studies 13, 125–128 (1998) (in Polish)
19. Tadeusiewicz, R.: Possibilities of Artificial Intelligence Methods Application in Computer Aided Teaching Systems. In: Conference Proceedings: 7th Polish-Wide Scientific Symposium Computer Techniques in Educational Transformation, Kraków, pp. 45–51 (1997) (in Polish)
20. Tadeusiewicz, R.: Cybernetic Model of Teaching Process Computerization for Permanent Education Aims. In: Proceedings of the IV Polish-Wide Conference Modern Technique in Culture, Science and Education, Tarnów, pp. 161–186 (1996)
21. Tadeusiewicz, R.: Computer in education system. Szkoła Zawodowa 5-6, 17–18 (1979) (in Polish)
22. Teaster, P., Blieszner, R.: Promises and pitfalls of the interactive television approach to teaching adult development and aging. Educational Gerontology 25(8), 741–754 (1999)
23. Valentine, D.: Distance Learning: Promises, Problems, and Possibilities. Online Journal of Distance Learning Administration 5(3) (2002)
24. Zaikin, O., Kusztina, E., Rozewski, P., Malachowski, B., Tadeusiewicz, R., Kusiak, J.: Polish experience in the didactical materials creation: the student involved in the learning/teaching process. In: Mahnic, V., Vilfan, B. (eds.) IT Innovation in Changing World, Proceedings of the 10th International Conference of European University Information Systems EUNIS, pp. 428–433. University of Ljubljana (2004)

Chapter 3
The Open and Distance Learning Processes Analysis

3.1 Introduction

High quality of virtual education cannot be simply achieved by means of using a better computer and faster networks. The bottleneck of the problem is in the question: what is the right division of roles, functions, tasks and competences between the IT utilities and the human teacher? Without answering this question we cannot expect both good economical and quality factors in the learning-teaching process.

Investigation of ODL processes has been performed based on the outcomes of the Socrates/Minerva eQuality project (2004-2006) [3]. The main project's activities were to produce:

- adequate methodology and tools for ODL Quality Process implementation, enabling educational organizations to offer and assess quality-based ODL services,
- a model applicable in each partners' country, within their own context, replicable elsewhere,
- re-usable documents and tools: General Quality Process Chart, implementation guidelines & tools, best practice database, a training package,
- pilot test training of 5 pilot teams to provide feedback upon methodology and tools,
- permanent link with normalization agencies,
- project results dissemination through networking and seminars, for professional feedback and validation.

The eQuality project associated the following institutions: European University Pole of Montpellier and Languedoc-Roussillon – coordinator (France), University Montpellier 2 (France), Open University of Catalonia (Spain), University of Tampere (Finland), West Pomeranian University of Technology in Szczecin, (Poland) University of Applied Sciences Valais, (Switzerland), Lausanne University (Switzerland).

The authors of this chapter present an approach to student life-cycle for ODL environment development. The student life-cycle is founded on a knowledge-based learning-teaching process. The authors believe that the emerging discussion

P. Różewski et al.: Intelligent Open Learning Systems, ISRL 22, pp. 51–74.
springerlink.com   © Springer-Verlag Berlin Heidelberg 2011

of the knowledge-based approach to the learning process needs association with the student life cycle model.

3.2 Model of the Learning-Teaching Process

In the traditional approach we treat the process of learning-teaching as an informal, continuous process that occurs in the mind of the teacher. In distance learning there is a need to distinguish individual elements, since analysis and modelling of each of them is connected to specified actions that we need to consider and undertake. Figure 3.1. depicts the schema of a referential model of the education process.

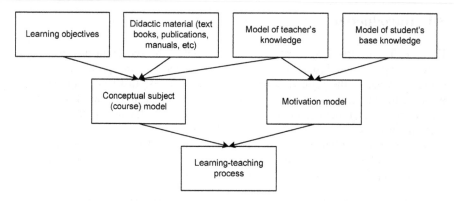

Fig. 3.1 Referential model of developing the education process

The entire process of learning-teaching can be presented as a set of connected models, the most important of which are:

-   Conceptual subject (course) model: the teacher develops the model basing on the existing textbooks, internet, and his/her own knowledge of the topic. This model does not depend on the individual characteristics of the student. In the process of creating the conceptual model of the subject the teacher is both the knowledge engineer and the subject's expert.
-   Model of student's base knowledge: this model defines basic knowledge of the student and can be presented as an ontological model. This gives the basis to assess the student after a completed learning cycle.
-   Motivation model: the teacher creates a motivation model for a certain group of students (preferably as small as possible), taking into account their social position, education, behavior. Analysis of this model allows for adapting the learning process to specific situations, e.g. by changing the learning goal. The motivation model can be a set of productive rules formed by the teacher or a methodology specialist.

3.3 Discussion of Basic Distance Learning Processes

Isolating basic distance learning processes encounters a lot of difficulties. It is connected to the diversity of forms of distance learning and ODL organizations which have distance learning in their offers. One point of view is to focus on knowledge processes, what was presented in table 3.1. Another approach is to divide processes from the point of view of educational organization's functioning. In such case, it is necessary to apply a certain generalization in order to easily transfer results between different educational organizations.

Table 3.1 Knowledge processes in the distance learning process (based on [8])

Knowledge processes	Actors	Descriptions	Formalization methods
Learning-teaching process	Teacher → student	Teacher transfer knowledge to student directly or uses distance learning methods. Teacher bases on his/her pedagogical skills and deep domain knowledge.	Ontology approach [12]
Research process	Staff → science	Researchers use synergetic research potential discover or redefined every aspects of science.	Innovation theory, linear innovation process
Expended process	Science → staff	A continues change of knowledge domain requires more study and learning by researcher.	Competency theory [11]
Discovery process	Reality → staff	The scientific problems mostly come from real life and surroundings world.	Soft computing method, system analysis

3.3.1 Processes and Roles Assigned to Them in ODL Organization

Authors, basing on the results of the e-Quality project, propose a generalized view on the educational organization, working in ODL, and processes connected to it. As a result of analysis of the ODL the following processes were identified:

- planning process;
- administration process;
- assessment and evaluation process;
- didactic material design and production process;
- student learning support process.

To each of the processes roles were assigned (tab.3.2.) to represent the level of competence that should be designated to perform each task.

Table 3.2 Basic processes in distance learning with the roles assigned to them (source [4])

Process name	Main roles	Detailed roles
Planning	Instructional designer	Scheduler Planner of the pedagogy Rule maker Instructional designer
	Content planner	-
Administration	Educational Administrator	Educational secretary Marketing responsible Institutional integrator
	Counsellor	Counsellor
	Coordinator	Project coordinator Technical coordinator Pedagogical coordinator Pedagogic director
	Technical administrator	Access controller Integrator
	Adviser	Mentor Pedagogical adviser Technical adviser Academic coordinator Administrative support External experts Lawyers
Evaluation	Evaluator	Evaluator
	Developer	Evaluator
Learning material production	Material designer	Content designer Graphic designer Usability expert Copyright specialist Information searcher
	Material producer	Author Content producer Tester
	Audio-visual specialist	Sound expert Layout expert Animation expert Video expert External experts
Student support	Pedagogical support	Motivator Initiator Group leader Interaction tutor Tutor Pedagogical supporter
	Technological support	Technologic support
	Tutor	Tutor
		(Counsellor)

(*) original terminology applied in the e-Quality project was maintained.

3.3.2 Comparison of Student Learning Support Process and Didactic Material Design

Analysis performed by the consortium of e-Quality project showed that processes of didactic material design and student support are the key ones when quality of the learning-teaching process is considered. Therefore in the further discussion focus should be placed mainly on these sub-processes (see Tab. 3.3.).

Table 3.3 Comparison of processes of student learning support and didactic material preparation (source [4])

Features	Learning material design process	Student's support process
Objectives	Product oriented	Customer oriented
Kind of production	Unique-demand production	Mass-demand production
Kind of parameters	Deterministic	Random
Modeling methods	Gantt chart	Simulation
Kind of criteria	Quality criterion	Efficient criterion

The goal of the process of didactic material design is to prepare materials (resources) aimed for learning a certain topic. In distance learning the process of preparing didactic material gains special meaning due to the already mentioned lack of direct contact with the teacher. Thanks to the didactic material, students discover knowledge by themselves, aiming at building appropriate (proper) thought structures in their minds. The process of didactic material design is usually oriented toward the product (usually intangible). Each created material is unique, prepared to fulfill the goal of the given course. It is characteristic that the order is serviced in a defined time schema. When analyzing the quality of the didactic material preparation process, we base on such quality characteristics as: usability (ergonomic aspect), competence (information aspect), and structure (cognitive aspect).

A different view on the idea of quality is represented by the student support process, which refers to such activities as help offered to the student with technical problems or pedagogical support. The support bases on hardware and personal infrastructure. As an example of infrastructure we can have telecommunication technical help systems (help-desk) or digital library of didactic materials. The discussed process is oriented towards the student (client), who applies to a given service at a random moment in time, and occupies it for a random period of time.

In such case, the key criterion is effectiveness of functioning, which can be calculated by analyzing the proper time and cost (e.g. total time when students demand access to resources). Examples of optimization for a closed queuing network consisting of the following servers (education services): teacher, course, administration and students, can be found in [13].

For further analysis of the discussed processes, the RUP (Rational Unified Process) methodology will be used. Originally RUP is used for planning and managing processes connected to building and implementing different types of software [5]. As the achievements of the e-Quality project show, RUP methodology can also be used for describing processes connected to distance learning, where the main stress is on the scope of competences of actors participating in the process. Such approach allows analysis of the learning-teaching process from the point of view of knowledge and resources (also intangible, e.g. specific skills) required for maintaining high quality.

The basic approach of RUP methodology assumes decomposition of the given process through a structure reflected by the following tuple: <role, activity, artefact, additional elements>, where:

- role – is represented by a set of mechanisms that define activities performed in order to fulfill responsibilities;
- activity – set of actions performed by a certain role, which are motivated by common goal (e.g. planning the structure of didactic material);
- artifacts – things created by men, an especially important artifact is information that is created, modified and used by individual processes;
- additional elements – support elements which do not take direct part in the process, e.g. guides, instructions.

Special attention has to be paid to differentiate between the statuses of a role from an actor. The actor is a person or an institution that has a specific position in the ODL process. On the other hand the role plays the function that can be figuratively compared to a "hat". The given actor can wear different hats during a project, thus can play different roles. The status of a role is more flexible than the actor, because the actor usually integrates several roles or several actors, who can play the same role. In the process of ODL defining the roles is made more difficult because of different education systems and cultural differences. The prepared set of roles (tables were placed in the following chapters) is a result of compromises and in most education systems requires adaptation to existing conditions.

A comment is also required for the structure of the pattern that will be used for describing individual processes. The pattern includes the following fields:

- input – defines with what input sources the given process will work.
- process description;
- output – the output effect of the process, always showing the value added in regard to the input;
- sources – defines what sources the process will use to transform input into output;
- procedure – defines practical actions that will be taken in the given process.

Additionally, for each of the mentioned processes a set of problems that should be paid attention too, since they are usually the reason for lack of success of the process, will be defined. In order to prevent such risk, for each process also a set of quality criteria and recommendations, the obeying of which increases the probability of the process well-functioning, is proposed.

3.3.3 Process of Designing Didactic Materials

The process of didactic material design can be divided into sub processes (tab 3.4.):

-   process of designing didactic material itself;
-   process of building didactic material.

Table 3.4 Main characteristic of didactic materials designing process (based on [4])

Process's name	Process of designing didactic material	Process of building didactic material
Outline	fig. 3.2	fig.3.3.
Possible problems	<ul><li>the didactic source does not correspond with the initial knowledge and student's abilities;</li><li>the chosen pedagogical approach is not effective;</li><li>student might be learning overdue knowledge;</li><li>some students may have problems (e.g. technological) with using didactic materials and thus might lose time for solving issues not connected with learning;</li><li>difficulties with re-applying the given didactic material.</li></ul>	<ul><li>the work schedule is not maintained and some parts of the didactic material are not made available to the student in the assumed time;</li><li>the budget is not respected what causes some parts of the didactic material to not be produced or be produced in worse than assumed quality;</li><li>proper norms and standards are not considered;</li><li>some students have difficulties with using the materials.</li></ul>
Quality criteria	<ul><li>effectiveness of the pedagogical approach;</li><li>issue of updating;</li><li>coherence of content;</li><li>usability for end-user;</li><li>reusability;</li><li>well developed order (scenario structure).</li></ul>	<ul><li>accordance with the schedule and budget;</li><li>coherence with documentation;</li><li>usability for the end-user (student).</li></ul>

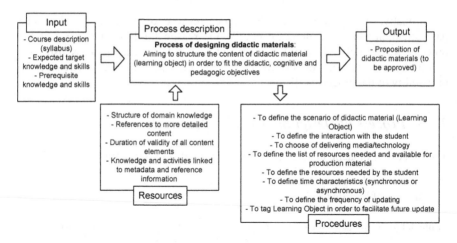

Fig. 3.2 Process of designing didactic materials (source [4])

In case of building didactic materials it is recommended to apply a coherent graphic pattern, preferably based on the styles technology. Moreover, it is recommended to use automatic systems of content generation, e.g. CMS (Content Management System). An important issue is to also place big stress on testing usability of the didactic materials both regarding the navigation and the structure of the material itself [9].

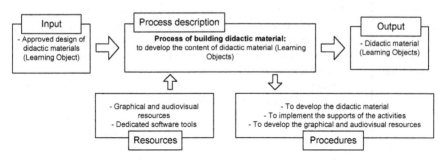

Fig. 3.3 Process of building didactic material (source [4])

The process of preparing didactic materials integrates several roles that appear in education. All actions are based mainly on the working environment provided by the Learning Content Management System (LCMS). Special stress is placed on cognitive conditionings and student's characteristics. Main characteristics that consist for competences of participants of the process of preparing didactic materials were presented in table 3.5.

In the tab. 3.6. the quality criteria for typical roles in the process of developing didactic material are defined.

Table 3.5 Analysis of the sub process of preparing didactic materials (source [4])

ROLES	ACTIVITY	ARTEFACTS & TOOLS	ADDITIONAL ELEMENTS
Author	To search information to be integrated in the learning materials To create contents based on the established goals and taking into account the ODL specificities To create learning activities To create educational resources in the frame of the course To reuse existing contents To provide references such as books, articles, websites… To decide the pedagogical choices with the rest of the staff involved, specially with the instructional designer To choose, together with the instructional designer, the modalities of interactions in the VLE and the pedagogical resources such as simulations, problem solving, cases studies…for the objectives accomplishment	Content management system Text editor Quotation software Documents to be used by the instructional designer and the rest of the team.	All kind of text software in order to reach all kind of contents. Databases, information sources, internet, browsers State of the art resources Libraries Correction software. Dictionaries. Encyclopaedias Legislation. Best practices. University system. Curriculum
Law specialist	To revise the conditions of use of the referenced documents in the materials To insert the requested legally symbols and documentation in copyright terms To implement the copyright protocol within the needed certification entity To revise the reused contents in legal terms To write the contract templates To elaborate the author contracts To negotiate the conditions of the author's contracts with the authors To prepare the individual contracts To reach agreements with third institutions in case an external content is going to be used To give advise on copyright issues to the people involved in producing documents and resources, or reusing the existing ones To manage the author's rights To manage the copyright issues inside and outside the institution	Legal forms Contracts Copyright memento Copyrights and intellectual property agreements	Latest laws, modifications in existing laws. Jurisprudence.
Graphic designer	To create the images To create the illustrations To create icons To create graphs To create the graphical chart (Create a unique graphic style) To create a graphical template for the different kinds o the course documents such as activities, contents,	Photograph editors Graphic editors Images databanks Scanner	Ergonomic principles Media resources Internet Graphical standards (SUN or Microsoft)

Table 5. *(continued)*

	extra resources… To insert the graphical resources already developed into the different interfaces of the digital course	Digital camera Web editors Media editors	Information on students' screen resolution
Audiovisual expert	To create locutions To insert locutions in the course To extract sounds from existing files (as pieces of songs) To insert sounds in the course To register environmental sounds To insert environmental sounds To create and insert music or any other sound To connect the sound media with the course interface To create and insert 3D resources To create and insert digital animations To create and insert video	Audio software Audio in different supports: cassette, CD, etc. Music instruments. Synthesizers. Animation software. 3D software.	Production software Images Sounds Internet Voices Songs
Developer	To insert the different media productions in the course's interfaces; such as text, video, animations, etc. To insert the different technical productions in the course; such as dhtml, applets, scripts, etc. To make the needed links in order to navigate through the course To test the different versions obtained in order to find errors, malfunctions, problems, etc… To respect the main e-learning standards (SCORM) To create intranets in the Learning Management System to allocate the course To create communication tools such as debates, forums, boards… To solve technical problems To control the state of development in the production stage anytime To create prototypes and a final version of the course To advice about the technology constraints to the rest of the multidisciplinary team	Integration and development software Media software Web editor Text editor FTP software	Learning Management Systems Didactic materials and learning resources Internet Computer languages
Instructional designer	To coordinate the production of the different learning objects, units and modules To establish the teaching and learning methodology To establish the learning objectives To work out the contents for the objective achievements and the estimated time of the course To supervise the content production considering the student characteristics To provide the student activities; individual or in group, that fit with the accepted methodology of work To provide the answers of the questions formulated To design the initial, continuous and final evaluations To coordinate the development of the complementary resources apart from the material To design the global storyboard of the course To produce the mock-up to dialog with authors To choose the media adapted to the content, the public and the rest of pedagogical elements To choose, together with the author, the modalities of interactions in the VLE and the pedagogical resources such as simulations, problem solving , cases studies…for the objectives accomplishment	Learning plans Methodological resources Mock-up of user interface A metadata grid for each reusable component of the resource Project management software	Complementary educational resources. Media resources Learning Management Systems Quality capabilities Legislation Best practices Innovative practices. Didactic materials A resource model with all the components, each component is tagged as reusable or

Table 5. *(continued)*

	To set up a plan and an implementation of a procedure of reusable components for further training actions To control the design assignments To integrate the content design and the graphic and navigation designs To link the course with the rest of the institution formative offer To coordinate and control the rest of the staff involved in the material design and production To adapt the course design to the pedagogical model of the institution To create a learning plan		not Web editor Theory about adult learning
Usability Expert (in certain institutions, these activities can be assumed by the instructional designer)	To do user studies To analyse the information obtained by the users To offer guidance about ergonomic decisions To contribute to the course structure To define the navigation system	Text editor Software for the quantitative and qualitative data analysis Software for rapid prototyping	Human perception capabilities Ergonomic knowledge (overall cognitive ergonomics) Social psychology Device knowledge Usability labs Methodologies of user study

Table 3.6 The quality criteria for typical roles in the process of developing didactic material (based on [4])

Role's name	Quality criteria
Didactic material designer	– ensuring availability of didactic materials; – limiting costs of using didactic materials; – preparing didactic material in required languages; – using appropriate and available tools; – structuring didactic material according to pedagogical goals; – proper development of didactic material.
Didactic material creator	– testing usability of the didactic material; – ensuring high degree of understanding the created didactic material by the recipient; – ensuring completeness of didactic material; – ensuring technological reliability of didactic material; – considering users' requirements regarding the didactic material; – ensuring an effective method of distributing didactic material.
Audio-visual designer	– proposing an intuitive interface; – creating a clear system of references; – development of functional audio-visual project of didactic materials

3.3.4 Student Learning Support Process

The student's learning support process consists of many inter-connected sub-processes. The most important ones of them are (see tab 3.7.):

- process of providing initial information (registration);
- process of preparing learning plan (incorporation);
- process of interacting with students (question answering).

Table 3.7 Main characteristic of the student learning support process (based on [4])

Process's name	Process of providing initial information	Process of preparing learning plan	Process of interacting with students
Outline	fig. 3.4.	fig. 3.5.	fig. 3.6.
Possible problems	– ambiguousness between technical and organizational requirements, and the student's status causes difficulties in proper formulation of advice; – division of information causes difficulties in building complete information; – people responsible for contact with students are not aware of the specificity of ODL and different requirements for each course, therefore they cannot provide credible information.	– unrealistic learning plan; – the student does not follow his/her learning plan.	– improper list of questions; – hesitating with answering the question.
Quality criteria	– credible information is available on the webpage; – people responsible for contact with students are prepared to work with ODL; – people responsible for contact with students have credible information about each ODL course.	– personalization of the learning plan with consideration of the flexibility of ODL; – usability of the learning plan in the context of motivation and proper organization of the student's time.	– using documentation of the interaction; – confirming pedagogical values of the given answer; – quality of the given answer; –

For the issues presented in such a way it is recommended to provide on-line tools that aim at testing initial knowledge of the candidate, as well as testing his/her equipment from the point of view of compatibility with the course requirements. It is also the goal to provide required information to the student through a set of web-pages.

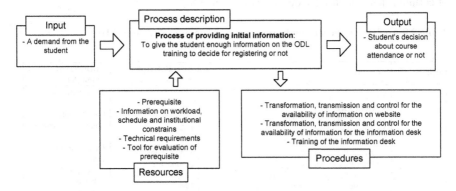

Fig. 3.4 Process of providing initial information (source [4])

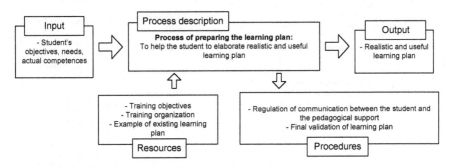

Fig. 3.5 Process of preparing the learning plan (source [4])

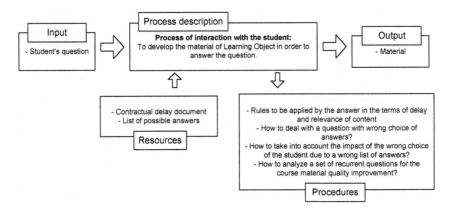

Fig. 3.6 Process of interaction with the student (source [4])

The recommended tool for managing student's progress is a prepared on-line system that would monitor it and provide information, what gives the possibility to control the degree to which the student is following the learning plan prepared for him/her.

Control and planning of the student learning support process requires presence of certain roles, which were described in table 3.8.

Table 3.8 Analysis of the student learning support sub process (source [4])

ROLES	ACTIVITY	ARTEFACTS & TOOLS	ADDITIONAL ELEMENTS
Teacher/ Counsellor	To create a calendar To create a schedule To create a synchronous contact timetable To define the rules of work To implement the rules of work To explain the learning methodology To familiarize students with the VLE To explain the singularities of e-learning To help students to elaborate a realistic & useful personalized learning plans To reach learning assignments with the students To activate and promote the participation of students To lead the learning dynamics To provide further information that the mere curriculum if it is needed To answer students' e-mail questions within a predefined time limit To use instruments for continuous assessment To give feedback from assessment as soon as possible To assure the final evaluation To motivate the learner in the accomplishment of the subject objectives	E-mail Virtual learning environment Chat Mailing list Calendar/schedule/ timetable ePortfolio tools Exam tools Assessment forms Groupware tools	The didactic material Complementary educational resources Links to libraries, dictionaries, encyclopedias, seminars, workshops... Bibliographies Performance guides Links for the students that have not expertise in the use of materials: for instance where it is possible to find manuals, guidelines, templates for self-learning. Guide for the student's time management
Head of the diploma	To give general advice on the methodology on following ODL To choose the staff involved in each course, considering their attending capabilities To choose staff considering their pedagogic capabilities To choose staff considering their expertise in the knowledge area To relieve teacher role if needed To answer students' questions within a predefined time limit To manage conflicts between students and tutors or teachers To control the training material delivery To establish the sequence of courses with the students To supervise the conduct of the people involved To receive professional offers from the enterprises To disseminate job opportunities among students To assess the impact of the training in employment terms	Learning plan, which is the document that defines the objectives, contents, activities, evaluation and schedule in a certain subject, brochures and practical documents to present the proposed methodology on ODL Tool to manage complaints	Monitoring guides for the following up of the student progress, course management guides Codes of conduct Expertise in the problem mediation's knowledge area

Table 3.8 *(continued)*

Tutor	To foster the students' involvement in the participation in the activities of the institution To manage conflicts between students To inform about events or projects, the institution is involved in To transfer the wrongly addressed questions to the right person (i.e. questions about administrative issues) Provide information about the followed learning methodology To attend the learner in a personalized way To answer questions about orientation, guidance or assessment To answer students' questions within a predefined time limit To make students feel part of the educational community To promotes interaction and cooperative work among the learners	Learning plan and also contents, activities, educational resources Tool to introduce marks E-mail Virtual learning environment The didactic material	Links to extra-academic sites (student integral education) Tools and links for professional guidance Tools for conflict resolutions
Administrative support	To provide to potential students relevant information on ODL specificities before registration To give information about the matriculation process To help potential learners during the registration To manage the registration To update the academic file of each student in the VLE To deliver didactic materials and resources To check whether students have received or not materials To deliver the diplomas to the students	Tools to manage 'matriculation aspects' e.g. administrative software Protection law on the storage of personal information Tool to manage communication with the learners Information document on ODL advantages and constraints. Documents about practical information to give to students just after registration on the courses, e.g. the communication tools, the way to contact the coordinator, the tutor, the counsellor, etc. Informative dossiers Demos about the teaching and learning process Administrative software, student register, calendar, curriculum	Budget tools Registration forms

Table 3.8 (*continued*)

Technologic support	To organise the VLE and other technological tools To answer technological questions within a predefined time limit To guidance student's for the technology acquiring To confirm the students equipment as enough for the course To attend and end up problems of the network, server and connections To solve hardware problems To solve software problems To attend technical problems within the VLE To prevent typical students' problems	Computer languages Telecommunications tools The requested hardware The requested software	Manuals in case a problem escape from their capabilities Software requirement Links to manuals, guidelines, handbooks, templates, etc. about the use of the requested technology
Hotline	To answer individual questions about the functioning of the VLE and all the problems of access, compatibility, etc. To create and maintain a FAQ To make regular review on the problems encountered by users and on possible solutions	Help desk protocol Virtual learning environment, e-mail, Skype, Net meeting…	Links to the latest version of the requested software
Access controller	To guarantee the protection of privacy and intellectual property To give access keys to the VLE To give access keys to the library To manage access To stop non-registered users from accessing to the student' productions in collaborative spaces	Network servers Content Management Software	Login requirement

Table 3.9 The quality criteria for typical roles in the process of developing didactic material (based on [4])

Role's name	Quality criteria
Pedagogical support	– providing the student with information regarding the course; – preparing a system for promoting student motivation; – effective organization of interaction (cooperation) between students; – proper choice of interaction modes; – acquainting students with technical aspects of the learning environment; – preparing the cooperation promoting system; – providing students with feedback information
Technical support	– providing students with information about the applied technology; – ensuring easy access to technical support; – ensuring the possibility of direct consultation regarding the applied technologies
Teacher	– ensuring students with access to the course; – providing consultation regarding the course; – applying appropriate technical tools for learning support; – supporting the learning of a course through provided tasks and exercises.

In the tab. 3.9. the quality criteria for typical roles in the process of student learning support are defined.

3.4 Knowledge-Based Approach to Learning-Teaching Process Design

Open and Distance Learning (ODL) is a sophisticated process running in certain (virtual) spaces including administrative, technological and knowledge management scopes [1]. The administrative (organizational) scope addresses the necessity for developing an efficient learning environment. The student requires an arranged curriculum enabling formulation of an effective domain knowledge system (both fundamental and operational). This is performed on the basis of labour market requirements, scientific research and an analysis of a given domain, the pedagogical experiences and routines inherent to a particular organization.

The technological scope is conditioned by didactical requirements of the distance learning environment which, on the contrary to the traditional form of learning, strongly relies on applied technological means (some examples can be found in [7]). The technical setup of the ODL learning environment involves delivering the learning materials and organizing the means of communication. From the didactic point of view, each learning session in an ODL course is an adequate media for transferring learning materials and the right technological tools have to be used in organizing the learning spaces. The didactical requirements are set by the teaching methods, accessibility, personalization level and providing complete student involvement. Choosing media types for learning materials and organization of learning spaces highly depend on the ODL course's subject, content, scope, level and working methods.

An important aspect of meeting the didactical requirements is to take into account the technological and supporting resource limitations in the organization, which also affect the kinds of media and learning space organizing tools that are used.

Another important facet of the aforementioned technological issue is the necessity for a high content personalization level specific to a given student (or a student contingent). Adaptation of the content is based on XML language and personalization mechanisms such as student profile and Learning Object (LO) [2,10,12].

3.4.1 Student's Life-Cycle as the Knowledge-Based Learning-Teaching Process

Regarding the knowledge-management point of view, the learning process is perceived as a form of intellectual resolution. It is believed that teaching a student is a multi-stage process of knowledge extraction, interpretation and transfer [6]. Basing the student's education process on the constructivism principle, certain level of student's competence is formed. The constructivist pedagogues comprehend knowledge as a meaningful structure which the learner constructs in a social interaction. The act of learning is a situation in which the individual accomplishes meaningful mental ensembles for him/herself and his/her fellows in a resourceful information environment.

In the learning which aims at construction of knowledge as meaningful structure, pedagogy is interspersed with the demand for renewal of functional collaboration between different units of universities and other scientific and educational institutions. In an even wider context, knowledge can be grasped as the basis for capabilities and talent, i.e. the art required for human action and meaningful life. The appropriation of factual contents of knowledge and their recollection are not enough for the success in different occupations, interactions and situations of life in the face of problems to be solved and in regard to the control gained over one's own life within the society. The renewal of the art by means of creation of new knowledge is also required. From the perspective of learning this means that knowledge which supports thinking must be combined with objectives and performance, i.e. the ability to act. The original source of learning is not an experience in itself, but a problem or a dead-end caused by the conventional modes of action. The result of learning is not reduced to mere problem solving, it also requires understanding the shortcomings of the previously obtained knowledge and the ways action is to be organised and controlled in the future. In the case of ICT in learning this requires that the focus must be at the same time on the contents of knowledge, and different problematic situations and impasses which have to be resolved.

It should be noted, that the presented discussion of education process is based on real conditions and facilities. The knowledge-based approach offers the possibility to examine the education process in the two: teaching and learning viewpoints. The main objective for an education institution is to merge the two subsequent processes of teaching and learning into one consistent knowledge-driven education process (labeled Knowledge-Based Learning Process – KBLP). This is believed to stipulate the learning process efficiency not only in the organizational context but also in the form of the learning organization, involving innovation processes' occurrences and their control within the learning space. Therefore, the real challenge is to enrich KBLP's abstract, knowledge-based scheme with parameters and characteristics of an existent system layout of the university.

Figure 3.7. presents the proposed scheme for modeling the student's intellectual progress (student's life cycle) as a knowledge-based process. The scheme represents the main layers in KBLP management. Five levels have been isolated to enable controlling and steerage knowledge transfer and acquisition. Knowledge characteristics vary depending on the stage of the learning process. In this proposal knowledge is not considered as a monolithic model, it is formed from a consistent set of models addressing and encompassing several education activity objectives.

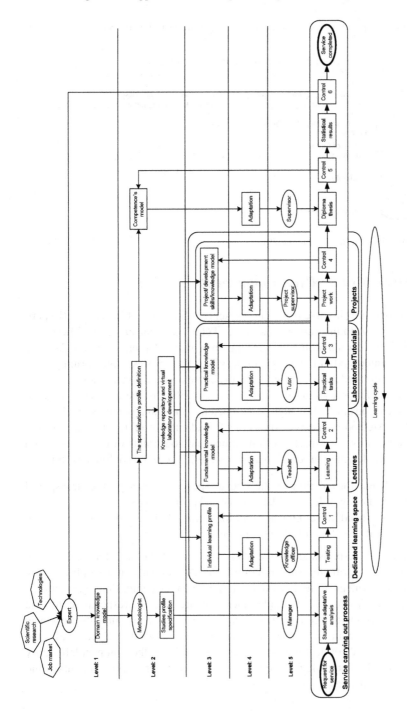

Fig. 3.7. Student's life-cycle as the knowledge-based learning-teaching process (based on [4])

The base components for the schema are presented on the fig. 3.8.

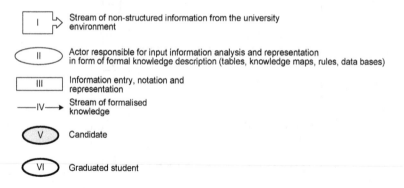

Fig. 3.8. Student's life-cycle nomenclature

3.4.2 Student's Life-Cycle Components

Several actors are involved in the presented Student's Life-Cycle (Expert, Methodologist, Administration, Teacher, Tutor and Supervisor). These are professionals of certain domains working in a particular organization. We are particularly interested in actors influencing the process of forming and consolidating the student's knowledge, in the context of the given scheme.

Expert a single, or a group of professors responsible for the university's public appearance and position. The highest competence indicator is represented by the Expert. The Methodologist is responsible for adaptation of the Domain Model (DM) developed by the expert into limitations and constraints of the organization (this denotes the learning process' internal quality). The task is defined as follows:

– on one hand, formulation of the course specialization's profile, transposed into real planning of lectures, laboratory and project courses,
– on the other hand, development of a Motivation Model (MM), transposed into a real education offer partial to the university.

Administration is responsible for the MM analysis (education offer) and, basing on this construct, forms the student contingent. Additionally, it also plays a key role in the deployment of the knowledge requirements meeting the organization's processes. The Teacher, Tutor and Supervisor are directly responsible for the courses. They adapt the educational situation to a group of students' instance, basing on their domain knowledge and pedagogical knowledge. Specifically, the Teacher concentrates on lectures – theoretical knowledge, the Tutor on laboratory courses – procedural knowledge acquisition, the Supervisor on project courses and final degree certificates.

3.4.3 Multilevel Management Structure

Let us specify the levels of the given scheme including the models and actors. Each level provides the definition of the actors' function.

Level 1: Expert's level: analysis and innovation support within the learning process
The expert in a given region develops a Domain Model (DM) by means of: analysis of the labor market situation, recognition, and identification of the current state of science and domain knowledge. The DM's main objective is to represent knowledge required at a requested time interval. The model's validity and timeliness is interpreted as the difference between knowledge gain inherent to the domain and knowledge model which is enclosed in the student's profile. This concludes a support in relation between the labor market, professional scientific knowledge and the university's intellectual responsiveness. The analysis of this set resolves into adaptation and continuous improvement of the domain knowledge model, producing innovation in each cycle of the learning process.

Level 2: strategy of the learning process planning
This level enables initial setting of the learning process' objectives. The Domain Model is considered as a reference model of student's knowledge and it determines adaptation parameters of the didactic materials for the student. The Motivation Model (MM) is a result of the methodologist's analysis of the Domain Model. MM indicates which students (specialists) will be prepared by the organization. Feedback with Level 1 ensures that the obtained edification shall secure their employment in the future.

On the strategic level, a particular organization is adapted to market requirements. The adaptation process is a result of the Domain Model interpretation in the context of organization's practical constraints, mission and strategic goals. This revolves around the specialization of the course profile. Next, the profile is transposed into a subject list which enables both the education process perspective (horizon) and the Competence Model (CM). The education process perspective is the background for the studies' plan at the organization (university). It is a result of the requirements addressing particular educative activities – modification and innovation within lecture, exercise, laboratory and project courses. This may be specified in the following sequence: course development, course modification, knowledge range alteration within the course type. The CM is a result of total knowledge required and specifies the requirements level of the final diploma project course.

Level 3: Didactic materials content formulation available on-line
At this level, using outlines and settings from level 2, Subject Board content is specified for each course included in the specialization profile. Each course is pervaded with knowledge, according to strategy prior to DM analysis. The Subject Board structure and representation mode in ODL and the traditional learning vary significantly, based on SCORM and the knowledge repository.

At levels 3, 4 and 5 the following components have been integrated: student contingent's formulation process and feedback loops (A, B, C, D) characterizing the real learning process within the contingent. The proposed approach includes divergence of the knowledge processed, according to the courses' subject and their specifics. Therefore, 4 learning process loops (feedback) have been outlined. Each of the loops is devoted to different purposes and distinct characteristics. These are the following:

A. student's base knowledge analysis; the Teacher validates the student's level of qualifications in a given course. The results become a vital parameter and criterion for the adaptation operation, which is used sequentially through the Learning Flow process. The testing process provides the student's knowledge range measurement within a particular domain. This is relevant to set the student's competence degree within the domain. Moreover, the student's knowledge quality may be investigated, for example by means of the concept's depth analysis and identification,

B. the stage of fundamental knowledge absorption (learning), represented by the lecture course; the teacher transfers abstract knowledge of the domain of discourse, enabling the student's to master and use abstract, and reason within the given knowledge system,

C. the procedural knowledge edification; the teaching is focused on software use,the comprehension of particular computer-supported simulation areas' (environments') functioning (the comprehension of the applied metaphors). The tutor supports the student to transform his/her fundamental knowledge into an actual computer program or an event required for executing the simulation,

D. the application of absorbed knowledge in a real situation; the aim of this stage is to apply the acquired knowledge (stages B, C) in a concrete, real event. The student is expected to classify the given task, which are to be solved, efficiently and skillfully refer to his/her own cognitive schemes, and he/she has to apply the appropriate tool. The process is finished by obtaining a results analysis and producing conclusions

Each of the loops refers to 3 levels: the development or choice of the Learning Object (LO) (level 3), the adaptation of a real LO (level 4), testing and validation of the test results (level 5). The final mechanism of the entire learning process validation is the process of diploma formation and development. The diploma thesis consists of knowledge acquired by the student during the entire learning process execution. Correspondingly, the actual domain model form has influence on the thesis shape. This provides opportunity for correction of conceivable deficiencies, in order to achieve the main goal: the student possesses knowledge strongly related to and meeting with the market requirements.

Level 4: Adaptation operation – preparation of student contingent

The adaptation operation provides the possibility to adapt the reference knowledge form comprising the models from level 3 into a real education situation. At this level, students are identified (considered as a group). The situation is altered by the creation of personalized selection of (perspective on) knowledge, based on the teacher's/tutor's/supervisor's knowledge. The knowledge is not cancelled, but the moment of its transfer and application order is positioned on a timeline.

Level 5: Implementation and application of the KBLP and didactic materials preparation

This level represents the implementation and application process of the composed learning process (Knowledge-Based Learning Process), resulting in making the didactic materials accessible. All the actors work in on-line consulting mode.

3.5   Conclusions

Introduction of new solutions and tools to the teaching process is an absolute necessity. In such situations, however, we have to appeal to all concerned to remember about the humanistic dimensions of education, while teachers should be concerned about the student's emotional and intellectual identity in the information society.

Student Life-Cycle in ODL environment should include and integrate two coherent points of view, indicating learning-teaching processes, and defining the knowledge-based structure of the education process in higher education institutions. Such integration takes into account constrains and cultural differences between students of Europe. Each education institution has a variety of possibilities to adapt to certain normalization policies of European Community, which is a common practice in most of commercial institutions. Apart from the development of knowledge repository, as part of the learning life-cycle, Knowledge-Based Learning Process supports the adaptation and individual classification of various education situations (classification prior to student characteristics).

References

1. Burgess, J.R.D., Russell, J.E.A.: The effectiveness of distance learning initiatives in organizations. Journal of Vocational Behavior 63(2), 289–303 (2003)
2. Downes, S.: Learning objects: Resource for Distance Education Worldwide. International Review of Research in Open and Distance Learning 2(1) (2001)
3. E-Quality: Quality implementation in open and distance learning in a multicultural European environment. Socrates/Minerva European Union Project 2003-2006 (2006), http://www.e-quality-eu.org
4. E-Quality : A Conceptual Model for ODL Quality Processes. Deliverable 2.2 from Project, Downloadable form (2006),
    http://www.e-quality-eu.org/deliverable_2p2.html
5. Henderson-Sellers, B., Collins, G., Dué, R., Graham, I.: A qualitative comparison of two processes for object-oriented software development. Information and Software Technology 43(12), 705–724 (2001)

6. Khalifa, M., Lam, R.: Web-Based Learning: Effects on Learning Process and Outcome. IEEE Transactions on Education 45(4), 350–356 (2002)
7. Knudsen, C., Naeve, A.: Presence Production in a Distributed Shared Virtual Environment for Exploring Mathematics. In: Sołdek, J., Pejaś, J. (eds.) Proceedings of the 8th International Conference on Advanced Computer Systems ACS 2001, pp. 149–161. Kluwer Academic, Boston (2002)
8. Kusztina, E., Zaikin, O., Różewski, P., Małachowski, B.: Cost estimation algorithm and decision-making model for curriculum modification in educational organization. European Journal of Operational Research 197(2), 752–763 (2009)
9. Patryn, A., Statkiewicz, M., Susłow, W.: The approach to optimization of didactic application for student- teacher support. Cognitivism in media and education 7(1-2), 48–65 (2003) (in Polish)
10. Wu, C.-H.: Building knowledge structures for online instructional/learning systems via knowledge elements interrelations. Expert Systems with Applications 26(3), 311–319 (2004)
11. Yu, P.L., Zhang, D.: A foundation for competence set analysis. Mathematical Social Sciences 20, 251–299 (1990)
12. Zaikin, O., Kusztina, E., Różewski, P.: Model and algorithm of the conceptual scheme formation for knowledge domain in distance learning. European Journal of Operational Research 175(3), 1379–1399 (2006)
13. Zaikin, O., Różewski, P.: The virtual laboratory for simulation problem: e-Quality. In: The Proceedings of 1st Conference Nowe Technologie W kształceniu Na Odległość, June 9-11, pp. 287–296. Koszalin – Osieki, Poland (2000) (in Pollish)

Chapter 4
System Approach to Information Open Learning System Design

4.1 Introduction

The concept of distance learning has already been studied for many years. As shown in [40], at the beginning distance learning systems were used primarily at universities. The next step in the evolution was using distance learning systems in organizations and supporting knowledge workers [16] and to become an important element of corporate information systems. Studies of [34] have shown that the success of solutions offering learning-on-demand opportunities is correlated with growing adaptation of distance learning systems in companies' environment through implementation of modern technologies and solutions.

One of the more important factors of human civilization is proper organization of education system. Using a system approach [30] has presented a contemporary education system, which is divided into components and prepares different education paths for varying target groups. Distance learning is achieved on the basis of new technologies combined with new civilization challenges identified in [23] functioning as the Open and Distance Learning (ODL). The idea of ODL is used as a systemized tool in countries of difficult social and economical situation (e.g. [8,19,22,24,28]). Other applications of ODL also take place in Europe, where it is a part of the Bologna Process. Its objective is to build a European Higher Education Area by making academic degree standards and quality assurance standards more comparable and compatible throughout Europe. Similarly as in different regions, ODL in Europe is a result of changes made in the initial period of computer technology and internet development [29].

An analysis of scientific literature, ongoing projects and currently available market products indicate clearly that distance learning is a new direction in the domain of information systems engineering. The literature, however, is devoid of a system-based approach to the distance learning phenomenon interpreted as a complex information system. The authors perform a soft system analysis based on the theory of hierarchical multilevel systems, which allows developing a model of the Open System of Distance Learning. The proposed model plays a role of a meta

P. Różewski et al.: Intelligent Open Learning Systems, ISRL 22, pp. 75–94.
springerlink.com   © Springer-Verlag Berlin Heidelberg 2011

information system for knowledge management for Open and Distance Learning. As a conclusion of the analysis hierarchical structure and functional scheme of Open System of Distance Learning are proposed.

4.2 Problem Statement

As an object of research, distance learning is consistent with the idea of the Open and Distance Learning (ODL), carried out by any type of institution, is a system. It possesses a specified goal and a complicated structure, which consists of several subsystems having their own sub-goals and cooperating in the name of a common global goal. In the farther part of the paper, when talking about ODL we will use the term Open System of Distance Learning (OSDL) meaning an information system for distance learning that joins the characteristics of the traditional understanding of the Distance Learning term and also the current understanding of that term.

The purpose of the studies presented in this paper is to develop an OSDL concept based on the results of system analysis and to outline the evolution process in the frames of the presented concept that is related to traditional problems of a distance learning organization, such as: preparing syllabuses, offering education services, developing didactic materials for the process of gaining competences based on a specified knowledge model, and creating a statistical evaluation of students passing from one learning step to another in the frames of the accepted education system.

The concept of information system (named ODSL) for knowledge management in Open and Distance Learning proposed by the authors will consist of a hierarchical structure of an information system dedicated to management of an education enterprise operating in ODL requirements and its functional scheme. The hierarchical structure defines the set of sub-systems and their scope of activity. The functional scheme consists of four nested management cycles and is described as a sequential process of knowledge processing.

The chapter covers following issues:

-   the soft system analysis approach used to analyze ODL as a complex system. Authors take advantage of the theory of hierarchical multilevel systems developed by [17].
-   the hierarchical structure of Open System of Distance Learning. Special attention is devoted to its sub-models: Learning Content Management Systems, Learning Management Systems and Strategic Management Systems.
-   the functional scheme of the four overlapping management cycles. Each cycle is discussed in detail.

4.3 Systemic Approach to Conducting an Analysis of Open System of Distance Learning

The basis for functioning of an information system aimed to organization management is an integrated model defining both the processes occurring within the organization and the interaction between the organization and its environment. There are many approaches to properly define each of the aspects of the organization's operations:

-   formal definition of the organization's structures (theory of organization),
-   determination of behavioral motivation of the participants of the processes conducted in organizations (games theory),
-   formalization of the organization's activity through applying modeling of the basic functions (systems theory).

Each of the aforementioned approaches offers a possibility of performing an exhaustive analysis of organization's activity, however, none of them provide the opportunity to consider all of the organization's processes in the context of its life cycle. Soft system approach makes it possible to distinguish the main processes, to describe their interactions and to find a methodological approach to creating an organization management system that assures the optimal conditions for organization's existence. The traditional, hard system approach does not apply well in that case, due to the complexity of the system in question. One of the ways to deal with this problem is to apply the soft system approach which, according to [36], plays an increasingly important role in modeling of systems that are based on knowledge processing. Soft modeling is the only approach enabling us to perform an analysis of complex social systems in which interaction occurs at the knowledge layer. Proofs supporting that assertion can be found in [9,21,33].

As a theoretical and methodological foundation for Open System of Distance Learning analysis we will adapt a traditional approach developed by [17] in his work called "Theory of Hierarchical, Multilevel Systems". In that piece, the authors introduce the idea of a multilevel hierarchical organization structure that considers different functioning aspects of a given organization. The concept of hierarchical multilevel analysis is popular in scientific literature, e.g. in modeling social systems [18] and multidimensional complex software systems [6]. Soft system analysis, according to that approach, classified the discussed system as a multilevel, multi-goal system. Analysis of those systems is initiated with defining the scope using conceptual framework which the system will be analyzed with, and by defining the strata dimension. In the next step, the hierarchical nature of the studied system will be considered using three parallel dimensions (see Fig. 4.1.): the layer of description or abstraction (specifying areas of abstract description of an organization), the layer of decision complexity (distinguishing the layers of decomposition of the problems the organization is faced with) and the organizational layer (deciding the order of decisions in the decision-making process).

Conducting abstraction operation leads to developing a hierarchical structure of Open System of Distance Learning which in turn allows distinguishing separate sub-systems, arranging them according to specified abstraction spheres and

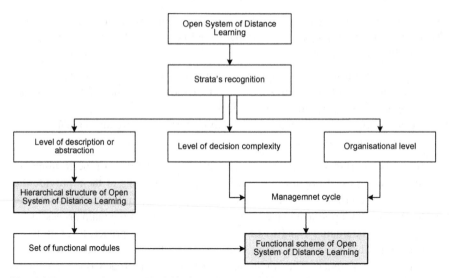

Fig. 4.1 Procedure for analysis of the Open System of Distance Learning

formulating both partial and main management criteria. The functioning scheme of the OSDL reflects the order and conditions of performing the modules and functions distinguished in the hierarchical scheme of the system, the nature of the identified cycles and their mutual influence.

In the next section the hierarchical structure of the Open System of Distance Learning and the functional scheme of the Open System of Distance Learning will be discussed in detail. The work of [15] will be used as basis for this discourse.

4.4   Hierarchical Structure of Open System of Distance Learning

The Open System of Distance Learning (OSDL) is being studied from the information system point of view, in order to create a management model that will enable both: further system development and controlling its every-day routine. The presented scope of OSDL operations allows focusing on OSDL system's dimensions which can be considered altogether using a common denominator of information system. OSDL is a system complicated enough to make building an isolated, one-dimension model very difficult. Therefore, like in every complex system, appropriate sub-systems should be distinguished.

Global OSDL systems operation criterion can be interpreted in the context of providing learning environment conditions which:

A.   Maximize fulfillment of individual student's requirements regarding time and learning mode.

B. Minimize differences with the traditional learning environment.
C. Maximize possibilities to gain a study results certificate.

A student studying in distance has certain needs regarding time and learning mode (aspect A of the global criterion). The most flexible way of solving this issue is developing a dedicated solution based on an information system including a knowledge base which will assure access to didactic content through Internet (at any time) [32,37,39]. However, in the didactic process we should also take into account the social aspect of learning, which assumes interaction with other individuals (process actors), like teacher (advice, consultation) or other students (group's discussion, joined projects). This necessitates developing a system that provides distance students with the best possible access to information resources of the system but also enables easy contact with other people engaged in the didactic process in the frames of OSDL. This problem is discussed in detail in terms of Distributed Interactive Learning in [10].

A reference model of each interaction in the frames of the created virtual space is functionally adequate and equivalent to the model of human interaction in reality. Therefore, the aim of every OSDL system is to reach for the ideal model of traditional learning (aspect B of the global criterion). The process of verbal information exchange is formalized at the level of knowledge modeling. Knowledge exchange that takes place between traditional learning process participants relayed through natural language is transposed to a limited operational environment of computer systems, where every form of communication is based on manipulating structures.

The purpose of OSDL systems existence in the context of the real-world education system is determined by a possibility of arranging certification of learning results (aspect B of global criterion). Placing education solutions in legislative reality forces finding appropriate legislations and legal solutions that allow OSDL units to certify studies on similar basis as it takes place in case of traditional education institutions.

The current state of knowledge and the characteristics of existing ODL organizations allow for determining the set of subsystems that OSDL system consists of (see Fig. 4.2) in the form of a hierarchy. Individual elements of the hierarchy are distinguished and ordered according to technological and scientific rules that characterize them. Each subsystem is described by a local objectives level, tightly connected with the global objectives level (global criterion) through common parameters. The distinguishable elements are not easy to formulate, because they are characterized by many different parameters of complex nature. Defining and modeling them requires knowledge of different domains, what leads to the level of functional modules.

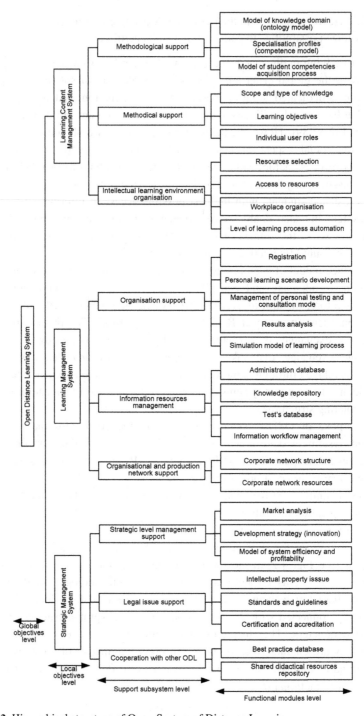

Fig. 4.2. Hierarchical structure of Open System of Distance Learning

4.4.1 Learning Content Management Systems

The purpose of the Learning Content Management Systems (LCMS) is to maximize conformity with the environment of knowledge exchange process. Resolving such problem requires analyzing different aspects of learning such as: educational, cultural, organizational, language (information and knowledge exchange) and the learning standards (national and international). The investigated LCMS subsystem criterion establishes point B of the global criterion – minimization of differences with traditional learning environment.

The main impulse that has lead to development of LCMS systems was creating a new approach to building didactic materials aimed at providing asynchronous learning. Distance learning courses are characterized by a set structure and a closed context. Research in the fields of pedagogy, teaching, psychology and cognitive science have allowed better understanding of the nature of knowledge in the learning process, enabling introduction of a modular method of didactic materials organization. As many scientific studies have shown over time, e.g. [1,5,27], knowledge has modular form consisting of concepts joined with relations. It is possible to divide every domain into modules using its structure. Basing on the modular structure of knowledge we divide didactic material into objects, each of which contains some portion of knowledge. It has been observed, that courses belonging to a certain class (e.g. class for higher education algebra courses) have common, or identical areas. This means, that for every class of courses there are portions of knowledge invariable in the entire spectrum of courses in that class. Due to that fact it is possible to design a module once and use it many times across different courses. Knowledge modules that the given domain is divided into are called Learning Objects. A Learning Object can be interpreted as an idea that allows for multiple usage of a single didactic material in different courses not only of one domain, but also of one paradigm. The presented idea of the Learning Object is the main mechanism used in the more and more important element of distance learning, meaning the personalization problem [35,38]. Kusztina et al. [13] shows that the new paradigm applied in e-learning systems assumes creating a knowledge repository containing a certain domain knowledge divided into knowledge objects (Learning Object). Such approach allows creating intelligent Web-based Education Systems that take into account the individual student learning requirements [3], as well as adaptive hypermedia systems which can be developed to accommodate a variety of individual differences, including learning style and cognitive style [31]. The idea of Learning Object, treated as reusable e-materials, requires also work in the knowledge management domain which allows developing a knowledge model adapted to e-learning [13,32,37].

Distance learning environment does not allow using natural language in its full extent as a tool for exchanging knowledge. Therefore, it is necessary to develop a dedicated communication language and related dialog scenarios. On the one hand, creating didactic materials based on the Learning Object requires a certain model of domain of knowledge. On another hand, the didactic material is formulated analogous to specialization profiles. Upon building a certain Learning Object we manipulate concepts to create a conceptual model of a given knowledge area according to a model of student competencies. In the best case possible, the created model is adapted to student's cognitive structures - and through the cognitive operations of assimilation and accommodation - added to the already existing knowledge structures.

One of the key elements of distance learning is orienting the material that is being developed towards a specific scope and type of knowledge. Knowledge as an object and purpose of learning can be divided at least into two basic types: fundamental-theoretical and operational. Fundamental knowledge reflects conceptual thinking, the basis for formulating new paradigms, problems, tasks, behavior rules, etc. Operational knowledge is necessary to realize scenarios, algorithms of performing operations. Situations that people encounter in every-day life are based on simultaneous usage of both types of knowledge in different proportions – depending on the level of task complexity. The learning objectives defines the relation between fundamental and operational knowledge in each case. In the context of educational challenges set by a given education goal, individual user roles are defined.

It is specific to the OSDL system to focus on learning space and organization of individual intellectual learning environment. In a traditional classroom, due to the direct interaction with a teacher, requirements towards didactic materials and textbooks are not always strict. In distance learning, the didactic materials and working environment influence the quality of learning greatly as they directly create the intellectual didactic flow. Upon designing intellectual workplace, we create a network of knowledge exchange space, select (software and hardware) resources preserving the system's purpose of use and deciding the level of learning process automation. An example of a learning space in the domain of mathematics (called CyberMath) was presented in [11] and it enabled interaction with mathematical models. Analysis of a training session in CyberMath has shown that the main quality factor is achieving a high level of integration with a properly arranged environment (i.e., maintaining high immersion factor). Furthermore, lack of access to the learning resources or deficiencies of the learning environment successfully hinder the possibility to achieve the objectives of the student learning process.

4.4.2 Learning Management Systems

The Learning Management Systems (LMS) class of systems allowed organizations to plan and track educational needs of their employees (students), partners and clients. Referring to [7] we claim that the LMS are a strategic solution meant for planning, delivering and managing all educational events within the educational organization, including both the virtual classes and the classes lead by an instructor. LMS systems enable registration and identification of students through the structure of profiles that describe individual characteristics of a given student and one's learning achievements. In that context, the process of personal learning scenario is feasible. The learning process is aided by tools for conducting analysis of results. Basing on the analysis of each course, the level of a teacher's workload, student's profile and the course requirements, performing a determination of individual testing and consultations modes is feasible. For example, a simulation model of the learning process can be seen as an important tool for strategic decision making related to learning-teaching process.

The core LMS' parameters are availability and capacity of the learning network and its cost. That connects the issue of Learning Management System with points A and C of the aforementioned global criterion. Increasing the quality of educational materials availability causes an increase in the work comfort and enables building a more efficient working space for the student.

An LMS system, due to the complexity of the distance learning process, requires an access to different types of data and information. The element that regulates the work of every organization is the administration database. Tests database and knowledge repository are unique to distance learning. Not also the tests database stores the tests contents, but also enables adapting a given test to a specific educational situation. It maintains that through a methodology of adapting tests to the given profile. The method of grading students' progress can be built on a set of developed heuristics or can be specified by set of regulation rules. The knowledge repository enables defining a common knowledge model [13,39]. The knowledge model contains a formal definition of concepts that can be used for modeling knowledge of the given domain and also the rules that allow creating real statements in the given domain. The knowledge repository can be built according to [20], on the basis of collecting unified ontologies in a form of a library. Due to large number of data and relayed information, it is essential to provide mechanisms of information workflow management in order to manage restricted resources (preserving the priorities of certain tasks).

4.4.3 Strategic Management System

Every OSDL organization operates in the educational market and can be characterized by a high level of competition that is additionally increased by creating the markets open for students. The class of Strategic Management Systems (SMS) helps in defining development strategies in the short-, middle- and long-term operational time frames. The strategic management module provides tools and methods for making decisions about creating a course-basket with consideration of potential students (clients), competitive environment and general social and consumer trends. The tools for marketing analysis provide the ODL organization with a possibility to develop organizational structure with accordance to new market conditions. It is created for maintaining a high level of competence of the graduates and maintaining a high level of the institution's competitiveness. Using the marketing analysis perspective and the adapted organization mission, the strategy development can be determined, which usually takes a form of a business plan. All discussions on the subject of a certain organization's future require   feasibility studies. To meet the needs of individual organizations, effectiveness and profitability models of the system are created in order to verify the premises made in the planning phase.

The university's activity, due to its educational mission and social value, should be strongly fastened in an existing legislation system. Respecting the laws of intellectual property in case when the didactic material is entirely digital, is a difficult challenge. A digital material can be copied easily and without quality loses. Therefore, technological and legislation mechanisms are needed to protect the didactic material. An example solution based on using an electronic license and public key structure was presented in [26]. Many OSDL system aspects due to social or law regulation require adhering to standards. They give possibility to participate in an environment based on the features of openness and interoperability.

Partnership work and resources exchange rely on mutual trust, which is developed on the basis of certification. Certification of distance learning centers can take place at different levels. At national level an appropriate institution visits individual education institutions, assessing the level of education and of staff preparation. An accreditation is granted on the basis of the commission's report. The procedure is the same for both conventional universities as well as for the ones offering distance education. For example, the key player in Europe is European Association for Quality Assurance in Higher Education (ENQA) whose goal is to define the rules of maintaining quality in education institutions of entire Europe. More information can be found in [4].

Organizations providing distance education cooperate with each other on different levels. The basis for common operation is starting a consortium of traditional institutions that would finance together a distance learning initiative. The next step would be creating a common best practice database and building a shared didactic materials repository. The best practice database allows efficient solving of problems that come with implementing new distance learning technologies, what often requires adapting complicated systems and modules, methodologies to local organization characteristics (the PROLEARN Virtual Competence Centre, is an example). That creates a need for exchanging "know-how" knowledge. The best practice database is implemented in the form of a knowledge repository with practical knowledge in mind. Experience of many OSDL institutions have shown that the most important element of ODL organization is the didactic material. Therefore, it is advisable to create cooperation space for ODL organizations, that will enable sharing materials within the framework of a didactic materials repository.

4.5　Functional Scheme of Open System of Distance Learning

Developing an information system dedicated to managing an education organization is a complex challenge [14]. In case of the OSDL, the analysis made by the authors lead to identification of four embedded management cycles, which differ in their time horizon. As can be seen on Fig. 4.3., each cycle includes a process that is being arranged by a certain decision maker. Within the time limit of each cycle, the system decision-maker compares knowledge areas in order to make decisions by estimating their content and depth (helpful algorithms have been proposed in [39].

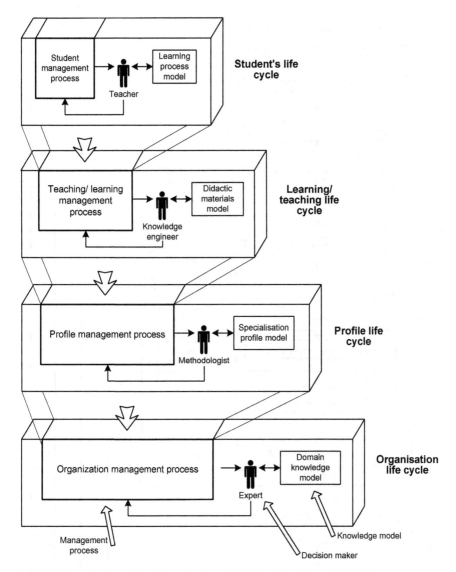

Fig. 4.3 Conceptual model of Open System of Distance Learning (OSDL)

Figure 4.4. shows the functioning scheme of an OSDL entity that consists of four inbuilt management cycles. The functioning scheme can be described as a process of sequential knowledge processing during:

- syllabus preparation,
- providing education services,
- developing didactic materials,
- acquiring competences based on a specified knowledge model,
- statistical evaluation of students' progress.

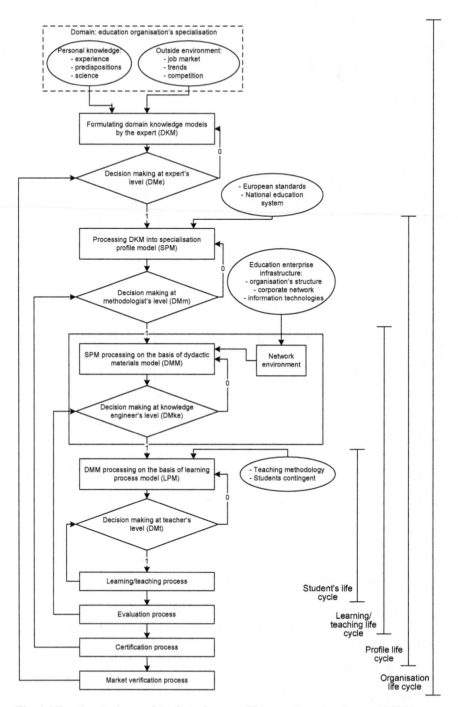

Fig. 4.4 Functional scheme of the Open System of Distance Learning System (OSDL)

4.5.1   The Knowledge Models

In each management cycle, the decision maker uses an appropriate knowledge model to make the decision. Between the knowledge models and education organization management subsystems located in the functional scheme (Fig. 4.4) the following relationships occur:

1) A **domain knowledge model** (*DKM*) relates to the strategic management system (SMS), its structure can be described with the following tuple:

$$DKM = \{Pr, R, Ac, Kd\},$$

where: Pr – production processes, R - roles, A - activities, Kd – domain knowledge.

DKM is formulated by an expert, on the basis of emerging market demands for new processes, technologies, enterprise organization forms and for establishing new roles and redefining tasks for the domain specialists. The expert focuses on the domain that the given organization specializes in.

2) A **specialization profile model** (*SPM*) relates to the given specialization syllabus management subsystem (LCMS system). The structure of this model can be described as follows:

$$SPM = \{ DKM, Ks, Sk, Ab\},$$

where: Ks – specialist's theoretical knowledge, Sk – practical skills, Ab - abilities.

MPS is formulated by a methodologist (e.g. education officials) within the aims and structures of specializations oriented on a given market area.

3) A **didactic materials model** (*DMM*) is formulated by a knowledge engineer on the basis of the specialization profile and the objectives and structure of the learning subject. DMM relates to the didactic material content management subsystem (LCMS system), for a given subject the DMM structure can be described with the following tuple:

$$DMM = \{ SPM, Sy, G, EC\},$$

where: Sy – syllabus and learning objectives of a given subject, G - hierarchical graph reflecting the subject's structure, EC – network environment constraints.

The knowledge engineer needs to have in mind the network environment's constraints that are also the technical constraints of the learning space.

4) **Learning process model** (*LPM*) relates to the learning management system (LMS). The model's structure can be described with the following tuple:

$$LPM = \{ DMM, ISy, LE, LS, CP\},$$

where: ISy – individual syllabus, LE – learning events, LS – learning objects sequence, CP – control points.

LPM is formulated by a teacher, who bases on the syllabus and learning of a given subject, the content of didactic materials and the evaluation of the initial knowledge of a given contingent of students.

4.5.2 Management Outlines

The structure of a typical outline for management in the Open System of Distance Learning (OSDL) is shown on Fig. 4.5. The outline corresponds to each management cycle. The main goal of the outline is to manage the process of adapting the descriptive knowledge to the normative counterpart type within the outline.

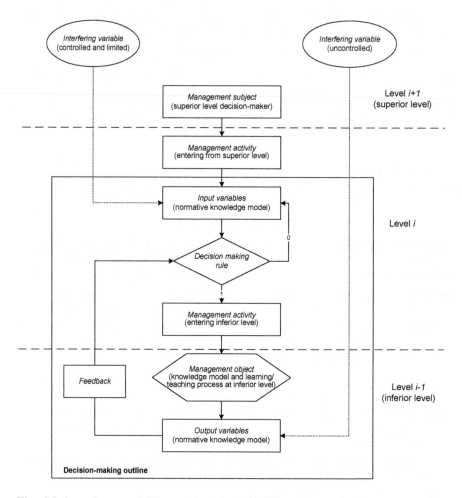

Fig. 4.5 Open System of Distance Learning (OSDL) structure of management outline (based on [12])

The decision-maker of level i can be described by the tuple $\overline{C}_i$:

$$\overline{C}_i = \{ MA_i, IV_i, F_i, PR_i, DF_i, MC_i \},$$

where: MA_i – management activity coming from top level, IV_i – constraints (interfering variables) coming from the exterior environment, F_i – feedback coming from the control process, PR_i – production rule, according to which the management activity for the bottom level is being formulated, $DF_i = MA_i + 1$ – decision function, according to which the content of management action entering inferior level is being estimated, MC_i – management cycle at level i, meaning time interval limiting the management action MA_i.

The diagram shown in Fig. 4.5. can be considered as a management outline (MO) with feedback F_i. The superior level decision-maker is the subject of management activities. The inferior level decision-maker together with the learning process are the objects of management. The central system-making element of MO is the decision-making circuit that compares the normative knowledge model (NK) with the descriptive knowledge one (DK). Normative knowledge model represents valid proper behavior rules, while descriptive knowledge model impartially and objectively describes reality [2,25]. Normative knowledge model is formulated by the system's decision-maker, on the basis of the management activity entering from the superior level. Descriptive knowledge model is specified by the inferior level decision-maker.

As was shown in [39], the knowledge representation model most suitable for the researched system is a hierarchical concept graph G^P. If methodology and algorithms from [39] are applied, knowledge models NK and DK can be presented in the form of hierarchical graphs G_{NK}^P and G_{DK}^P. With this approach, decision rule PR_i, on the basis of which decisions are made, can be described with the Kronecker symbol:

$$PR_i = \begin{cases} 1, \text{ if } G_{NK}^P \supseteq G_{DK}^P \text{ and } HD_i \leq T_i \leq HD_{i+1} \\ 0, \text{ otherwise} \end{cases},$$

where: G_{NK}^P - hierarchical graph of normative knowledge model, G_{DK}^P - hierarchical graph of descriptive knowledge model, HD_i - decision time horizon at level i, HD_{i+1} - decision time horizon at inferior level $i+1$.

Each production rule PR_i consists of two conditions:

1. Normative knowledge graph G_{NK}^P covers descriptive knowledge graph G_{DK}^P

$$G_{NK}^P \supseteq G_{DK}^P$$

2. Decision-making period T_i is longer than the decision horizon HD_i and shorter than decision horizon HD_{i+1}

$$HD_i \leq T_i \leq HD_{i+1}$$

As can be seen on Fig. 4.5, when both conditions are met the decision-maker formulates a management action for the inferior level (*i-1*), on the basis of the decision function DF_i. In this case, the decision function DF_i changes the normative knowledge model NK_{i+1} of the inferior level.

4.5.3 The Content of Management System

The approach described above is applied in every cycle of education organization management. Let us define the content of each management cycle presented on Fig. 4.4.

The education organization's life cycle

Subsystem of education organization strategic management aims at maintaining a high position of organization's graduates in the job market. The expert's decision model has the following form:

$$DM_e = \{MA_e, IV_e, SE_e, PR_e, DF_e, MC_e\},$$

where: $MA_e = \varnothing$ – management activity, IV_e – market demand for the specialization, SE_e – periodical graduates control in order to estimate their satisfaction in reference to market needs, DF_e – decision function (creating new specialization, modifying an existing one), LC_e – organization life cycle, PR_e – expert's production rule in the following form:

$$PR_e = \begin{cases} 1, \text{ if } G_{NKe}^P \supseteq G_{DKg}^P \text{ and } HD_e \leq T \leq HD_m \\ 0, \text{ otherwise} \end{cases},$$

where: G_{NKe}^P - hierarchical graph of normative (domain) expert knowledge, G_{DKg}^P - hierarchical graph of graduate's descriptive knowledge, HD_e – expert's decision horizon, HD_m – methodologist's decision horizon.

The profile life cycle

It is a subsystem of managing the process of adaptation of competency to a student's profile. The methodologist's decision model has the following form:

$$DM_m = \{MA_m, IV_m, F_m, PR_m, DF_m, LC_m\},$$

where: $MA_m = DF_e$ – expert's management activity (creating new specialization, modifying an existing one), IV_m – changes in European and national standards, changes in education system, F_m – statistical data from certification process used for comparing student's knowledge with specialization's profile, DF_m – methodologist's decision function (changes in specialization profile, changes in didactic materials, etc.) LC_m – learning-teaching process life cycle, PR_m – methodologist's production rule in the following form:

$$PR_m = \begin{cases} 1, \text{ if } G_{NKm}^P \supseteq G_{DKc}^P \text{ and } HD_m \leq T \leq HD_{ke} \\ 0, \text{ otherwise} \end{cases},$$

where: G_{NKm}^P - hierarchical graph of normative knowledge of the methodologist (the knowledge needed during creation of the specialization profile), G_{DKc}^P - hierarchical graph of student's descriptive knowledge from the certification process, HD_m – methodologist's decision horizon, HD_{ke} – knowledge engineer's decision horizon.

The learning-teaching life cycle

It is a subsystem providing an intelligent networked space to the learning-teaching system (i.e., effective usage of network environment and developing or adapting knowledge repository to students' profiles). The knowledge engineer's decision model has the following form:

$$DM_{ke}=\{MA_{ke}, IV_{ke}, F_{ke}, PR_{ke}, DF_{ke}, LC_{ke}\},$$

where: $MA_{ke}=DF_m$ – methodologist's management activity (changes in specialization profile, changes in didactic materials), IV_{ke} – organizational structure, corporate network, software and technical resources, F_{ke} – statistical data from student's evaluation process, DF_{ke} – knowledge engineer decision function (changes in didactic materials, changes in learning-teaching methodology), LC_{ke} – student's life cycle, PR_{ke} – knowledge engineers production rule in the following form:

$$PR_{ke} = \begin{cases} 1, if\ G^P_{NKd} \supseteq G^P_{DKs}\ and\ HD_{ke} \leq T \leq HD_t \\ 0,\ otherwise \end{cases},$$

where: G^P_{NKd} - hierarchical graph of normative knowledge at the didactic level (knowledge included in didactic materials), G^P_{DKs} - hierarchical graph of student's descriptive knowledge within the subject of learning, HD_{ke} – knowledge engineer's decision horizon, HD_t – teacher's decision horizon.

The student's life cycle

It is a subsystem that provides means for following and monitoring administrative correctness of the learning process, and evaluating the competence-increase process, given the knowledge model and the learning-teaching system. The teacher's decision model has the following form:

$$DMt=\{MA_t, IV_t, F_t, PR_t, DF_t, LC_t\},$$

where: $MA_t =DF_{ke}$ – management activity of the knowledge engineer (changes of didactic materials, and learning-teaching methodology), IV_t – contingent of students, F_t – students' grades during learning process, DF_t – teacher's decision function (forming students groups), LC_t – subject's learning cycle, PR_t – teachers production rule in the following form:

$$PR_t = \begin{cases} 1, if\ G^P_{NKt} \supseteq G^P_{DKe}\ and\ O_e \leq T \leq HD_{ke} \\ 0,\ otherwise \end{cases},$$

where: G^P_{NKt} - hierarchical graph of normative knowledge at the teacher's level, G^P_{DKe} - hierarchical graph of student's descriptive knowledge from the examining process, O_e - examining period, HD_{ke} – knowledge engineer's decision horizon.

4.6 Conclusions

The OSDL as a management object is a complex system which integrates processes of different nature. Using the multilevel hierarchical systems theory appears to be the appropriate way to analyze such systems. As the result of such analysis the following findings have merged: the hierarchical structure of the information system dedicated to managing an educational enterprise in OSDL conditions and its functional scheme. The hierarchical structure of that information system defines the set of its subsystems and their functional scopes. The functional schema is made up of four inbuilt management cycles and is described as a sequential process of knowledge processing. It is the role of the system's decision maker to manage the process during each cycle, on the basis of defined knowledge and appropriate rules.

It has been proven in the functional scheme that OSDL is a multilevel system of knowledge processing. However, for each management level only the content of normative and descriptive knowledge has been defined. The problem of choosing knowledge representation methods depending on the type of decision to be made has been also omitted. For example, in the SMS sub-system, in order to evaluate the ability of an educational organization to introduce new specialization profiles, a frame-based model may be used to represent knowledge regarding required resources (staff, information, computer, network). The main argument for such knowledge representation is that each resource has its own structure as well as evaluation and planning procedures. In the LMS sub-system, the most useful method of representing knowledge to form a personalized schedule of a student would be the rule-based model. It allows linking facts describing the initial state of student's knowledge with conditions of one's progress of study. In the LCMS sub-system, the application of object knowledge (serving as a basis for collecting laboratory classes results (e.g. simulation experiments)) is required with exclusion of the proposed approach to describing theoretical and procedural knowledge.

References

1. Anderson, J.R.: Cognitive Psychology and Its Implications, 5th edn. Worth Publishing, New York (2000)
2. Broens, R., De Vries, M.J.: Classifying technological knowledge for presentation to mechanical engineering designers. Design Studies 24(5), 457–471 (2003)
3. Canales, C., Peña, A., Peredo, R., Sossa, H., Gutiérrez, A.: Adaptive and intelligent web based education system: Towards an integral architecture and framework. Expert Systems with Applications 33(4), 1076–1089 (2007)
4. Ehlers, U., Pawlowski, J.M.: Handbook on Quality and Standardisation in E-Learning. Springer, Heidelberg (2006)
5. Gobet, F., Simon, H.A.: Templates in chess memory: A mechanism for recalling several boards. Cognitive Psychology 31(1), 1–40 (1996)

6. Gómez, T., González, M., Luque, M., Miguel, F., Ruiz, F.: Multiple objectives decomposition–coordination methods for hierarchical organizations. European Journal of Operational Research 133(2), 323–341 (2001)
7. Greenberg, L.: LMS and LSMS: What's the Difference? Learning Circuits - ASTD's Online Magazine All About E-Learning (2002),
    http://www.learningcircuits.com/2002/dec2002/Greenberg.htm
8. Hope, A., Butcher, B., Visser, L.: Distance education in Hong Kong. Quarterly Review of Distance Education 6(3), 207–215 (2005)
9. Jackson, M.C.: Critical systems thinking and practice. European Journal of Operational Research 128(2), 233–244 (2001)
10. Khalifa, M., Lam, R.: Web-Based Learning: Effects on Learning Process and Outcome. IEEE Transactions on Education 45(4), 350–356 (2002)
11. Knudsen, C., Naeve, A.: Presence Production in a Distributed Shared Virtual Environment for Exploring Mathematics. In: Sołdek, J., Pejaś, J. (eds.) Proceedings of the 8th International Conference on Advanced Computer Systems ACS 2001, pp. 149–161. Kluwer Academic, Boston (2002)
12. Kusztina, E., Zaikin, O., Różewski, P., Małachowski, B.: Cost estimation algorithm and decision-making model for curriculum modification in educational organization. European Journal of Operational Research 197(2), 752–763 (2009)
13. Kusztina, E., Zaikin, O., Różewski, P.: On the knowledge repository design and management in E-Learning. In: Lu, J., Ruan, D., Zhang, G. (eds.) E-Service Intelligence: Methodologies, Technologies and applications. SCI, vol. 37, pp. 497–517. Springer, Heidelberg (2007)
14. Kusztina, E., Zaikin, O., Różewski, P., Tadeusiewicz, R.: Competency framework in Open and Distance Learning. In: Proceedings of the 12th Conference of European University Information Systems EUNIS 2006, Tartu, Estonia, pp. 186–193 (2006)
15. Kusztina, E.: Conception of Open Information System for Distance Learning, Szczecin University of Technology, Faculty of Computer Science and Information Systems, book (2006) (in Polish)
16. Marwick, A.D.: Knowledge management technology. IBM System Journal 40(4), 814–830 (2001)
17. Mesarovic, M.D., Macko, D., Takahara, Y.: Theory of Hierarchical, Multilevel Systems. Academic Press, New York (1970)
18. Miklashevich, I.A., Barkaline, V.: Mathematical representations of the dynamics of social system: I. General description. Chaos, Solitons and Fractals 23(1), 195–206 (2005)
19. Mukerji, S., Tripathi, P.: Academic Program Life Cycle: A Redefined Approach to Understanding Market Demands. Journal of Distance Education 19(2), 14–27 (2004)
20. Neches, R., Fikes, R., Finin, T., Gruber, T., Patil, R., Senator, T., Swartout, W.R.: Enabling technology for knowledge sharing. AI Magazine 12(3), 36–56 (1991)
21. Nakamori, Y., Sawaragi, Y.: Complex systems analysis and environmental modeling. European Journal of Operational Research 122(2), 179–189 (2000)
22. Ojo, D.O., Olakulehin, F.K.: Attitudes and Perceptions of Students to Open and Distance Learning in Nigeria. The International Review of Research in Open and Distance Learning 7(1) (2006)
23. Perraton, H.: Open and Distance Learning in the Developing World. Routledge, London (2000)
24. Rao, S.S.: Distance education and the role of IT in India. The Electronic Library 24(2), 225–236 (2006)

25. Ropohl, G.: Knowledge types in technology. International Journal of Technology and Design Education 7(1-2), 65–72 (1997)
26. Santosa, O.A., Ramosb, F.M.S.: Proposal of a framework for Internet based licensing of learning objects. Computers & Education 42(3), 227–242 (2004)
27. Simon, H.A.: How big is a chunk? Science 183(4124), 482–488 (1974)
28. Suanpang, P., Petocz, P.: E-Learning in Thailand: An Analysis and Case Study. International Journal on ELearning. Norfolk 5(3), 415–438 (2006)
29. Tait, A.: Open and Distance Learning Policy in the European Union 1985-1995. Higher Education Policy 9(3), 221–238 (1996)
30. Tavares, L.V.: On the development of educational policies. European Journal of Operational Research 82(3), 409–421 (1995)
31. Triantafillou, E., Pomportsis, A., Demetriadis, S.: The design and the formative evaluation of an adaptive educational system based on cognitive styles. Computers & Education 41(1), 87–103 (2003)
32. Valderrama, R.P., Ocaña, L.B., Sheremetov, L.B.: Development of intelligent reusable learning objects for web-based education systems. Expert Systems with Applications 26(3), 273–283 (2005)
33. Wang, P.P.: Soft modeling for a certain class of intelligent and complex systems. Information Sciences 123(1-2), 149–159 (2000)
34. Wang, Y.-S., Wang, S.-Y., Shee, D.: Measuring e-learning systems success in an organizational context: Scale development and validation. Computers in Human Behavior 23(4), 1792–1808 (2007)
35. Wang, H.-C., Hsu, C.-W.: Teaching-Material Design Center: An ontology-based system for customizing reusable e-materials. Computers & Education 46(4), 458–470 (2006)
36. Wierzbicki, A.P.: Modelling as a way of organising knowledge. European Journal of Operational Research 176(1), 610–635 (2006)
37. Wu, C.-H.: Building knowledge structures for online instructional/learning systems via knowledge elements interrelations. Expert Systems with Applications 26(3), 311–319 (2004)
38. Xu, D., Wang, H., Wang, M.: A conceptual model of personalized virtual learning environments. Expert Systems with Applications 29(3), 525–534 (2005)
39. Zaikin, O., Kusztina, E., Różewski, P.: Model and algorithm of the conceptual scheme formation for knowledge domain in distance learning. European Journal of Operational Research 175(3), 1379–1399 (2006)
40. Zhang, D., Nunamaker, J.F.: Powering E-Learning In the New Millennium: An Overview of E-Learning and Enabling Technology. Information Systems Frontiers 5(2), 207–218 (2003)

Part II
The Problem of Knowledge Modeling in Open and Distance Learning

Chapter 5
Ontology Approach to Knowledge Modeling

5.1 Introduction

The last decade of the 20th century, sometimes called the time of the Information Revolution, has a decisive influence on how we live and work today. Development of such domains as computer science and telecommunication is the source of formation of economy based on knowledge. Knowledge begins to play the role of a strategic factor, often more important than land, financial capital or natural resources considered the „motors" of the Industrial Age. The changes that arise on this background can be characterized in the form of three basic development mega-trends:

- technical megatrend of digital integration – considers techniques, media, networks and ways of transferring and processing information;
- intellectual megatrend of intellectual challenges – connected to the transition from mechanical perception of the world as a big revolving machine with inevitability of rotation of the Celestial Spheres to a system of chaotic perception of the world as a complex system in which chaotic behaviors occur;
- social megatrend of changes and creation of new jobs – connected to dematerialization of work due to the use of information techniques as well as increasing educational needs and the necessity of continuous learning.

From the point of view of the information civilization, these megatrends are connected to finding new ways and methods of acquiring, organizing and spreading knowledge. The problem of organizing resources and access to them becomes serious when this sets quickly, significantly and in an unconstraint way increase their capacity. Conventional access tools in the form of hierarchically organized folders, available access paths or search engines, fulfill their function perfectly when the number of positions is not high. However, the expansional increase of resources results in them not properly representing and covering the required domain of knowledge, and their quality not meeting the needs of users. Solving this problem requires a tool/method that allows for describing and reflecting the philosophical, scientific and technical state of the given knowledge domain.

P. Różewski et al.: Intelligent Open Learning Systems, ISRL 22, pp. 97–119.
springerlink.com   © Springer-Verlag Berlin Heidelberg 2011

5.2   Analysis of Information to Knowledge Transformation

The key factor in discussing every knowledge model is understanding of the very essence of the operations of abstractions. According to [7], we can identify two forms of the operations of abstractions - aggregation and generalization. [32] have defined a source meaning for the term of abstraction a process in which some details, traits, or properties are intentionally omitted. Concealing of the details allows us to concentrate on general and key traits of the analyzed object. The process of abstracting - a philosophical concept defined in [23] - is concerned with forming an idea based on repetitive renditions of certain qualities, or by abstracting determining factors of a given subject. Terms such as "elements" are ignored in this case because we consider here a particular application, or an inclination of an operating analyst. In that context, the purpose of entire operation is sharing/demonstrating of details/traits/properties that are interconnected with the given application situation that is significant for a chosen model/process at the same time ignoring the remaining unimportant traits.

We often face the situation where analysis of reality exposes an overabundant amount of details for a single abstraction to be applied effectively. In that case, the details are decomposed into certain hierarchy of levels and groups. Every level, or group corresponds to some assumed granularity. That interpretation of abstraction it reasonable to relate it to semantic memory, for example defined by [1] or [26], that posses the quality of being hierarchical. It is also possible to assume that upon creating a hierarchy by means of analyzing semantic features, a model is built using the abstraction operations. For example, that can be the case upon concept taxonomy creation process.

An abstraction is exemplified by two fundamental operations: aggregation and generalization. According to [7], the operation of aggregation is relevant to the form of abstraction in which the dependencies between connected objects   is delegated onto an object in located in higher level of the object hierarchy. Upon performing the aggregation a multitude of dependency properties is ignored, whereas a select group of properties are aggregated into a single cluster. The aggregation, defined by [36], is a form of abstraction resulting in a superlative object based on the properties of the constituent elements.

According to [7], the operation of generalization can be defined as an abstraction in which a collection of similar objects is formed into a general object. Upon performing generalization failing to uphold individual differences among included objects is possible. [36] consider generalization as an operation of enumerating the collection of concretes or types and mapping onto a single object of general type. In addition, generalization, formed according to the concrete-to-general-type pattern, is often called as classification, contrary to the generalization of many types into single one. [32] define generalization as an abstraction that enables considering a class of singleton objects as a general, single and well defined object. The authors of the latter sources also strongly indicate the importance of generalization referring, inter alia, the application of that abstraction on day-to-day basis, for example upon skills acquisition, or using natural language. In that sense we may claim that every individual initiates its activity with observation of concrete instances that leads to absorbing new knowledge about certain class that is being exemplified by the observed instances. Hence, generalization preserves inheritance resulting in descendant objects inheriting features and properties of their ancestors.

5.2.1 The Dependency between a Knowledge Source, Learning Environment and Knowledge Representation Language

Speaking form the perspective of automation of processes such as information exchange and acquisition of determined volume of knowledge upon ongoing learning process, the concepts such as internal knowledge representation and encoding cognitive language, or gestalt and others are usually omitted in a discussion. Every participant of the "learning-teaching" sequence (process) is regarded either as an emitter, or a receiver of a transmitted information and as an operator acting in a cumulative information broadcast space. In that sense, if we resort to the "input - black box - output" approach where didactic material and tutor's instructional skills constitute the input element, the black box component represents unformalized individual's interior processes of knowledge assimilation and assumption, and the output is the mastered knowledge. In the context of the information science-based approach, the model of learning should offer an explanation of the dependency between information transmitted by the tutor and knowledge absorbed by the student.

A critical requirement for determining the dependencies of transformation of the input to the output is applying and identical language for both elements. In conventional learning this requirement is preserved through use of natural language. Thanks to that, the tutor can directly control the output (for example by dialogue) and on the executive level, also control in real time the execution of the tutoring process with accordance to predefined criterion that can at best be only partially formalized due to its individual and unique quality. Final statistical processing of quantifiable assessments of student's knowledge provides possibilities for corrections in long- and short-term curricula. On the operational level, the tutor, thanks to one's tutoring techniques, reinforces semantic significance of the input information such as tutoring materials, exercises and didactic materials. It attains that through adaptive strategy of selecting and reordering and altering the depth of analysis of presented concept/phenomenon, the chosen metaphors, expansion of associative links between concepts, or by altering the associations of the investigated concept with different contexts. The requirement of maintaining an unequivocal character of instruction is also taken into account during that process, also on operational level upon communicating with students.

Both in distance and conventional learning the planning activities both on long-term and short-term basis are similar in substance. The significant difference emerges on the operational level of plan implementation, namely upon relaying a full scope of knowledge from the tutor to the student. The reason for that is not only a result of lack of direct interaction, but a different learning space becomes a factor since transfer of traditional methods of instruction is not possible here.

In traditional learning, the basic means for knowledge representation and relay is natural language. Artificial symbolic languages of different kind and media-rich facilities are used as additional outlets for reinforcing the semantics of knowledge relay performed in natural language, both upon lecturing, or upon preparing the didactic materials. However, taking into account numerous ergonomic and psychological factors, a computer-aided learning environment imposes plenty of limitations on use of natural language in the same quantity and quality as necessitated by the learning process and the corresponding process of operational control. On the other hand, the very same type of environment offers virtually unparalleled opportunities for use of symbolic/artificial languages and media-rich formats of didactic materials. Therefore, it is reasonable to pose the following question: how and using what is it possible to translate the qualities of natural language and natural language-driven instruction into symbolic languages? The unlimited capacities of computer technologies offer further expansion of the semantic features through intensification of input stimuli and that corresponds with effectiveness of minimizing of vagueness of instruction.

Regardless of the learning process organization, the tutor is not a sole source of knowledge. An equivalent role are is by books, media outlets and learning individuals' milieu that can be characterized by communication performed overwhelmingly in natural language. Hence, upon creating a knowledge representation and relay language it is necessary to have it correspond to natural language and knowledge representation and relay language that are characteristic to other types of knowledge sources. In other words, it is necessary for new knowledge absorbed from outside of learning environment to be effectively merged with knowledge of the student. That is particularly true for conceptual type of knowledge.

5.2.2 Structure as an Instrument for Performing Analysis of Information Contents

In the process of learning, the general process of communication and the concept of information are inextricably linked with the concept of structure. Nowadays, a human individual is overwhelmed by abundance and ubiquity of information. In that context, the pending research issue is determining the volume and amount of information that is perceived and can be utilized by its recipient. Therefore, a structure becomes a tool used upon analyzing information contents and supporting comprehension of information substance and solely by means of structure any individual is capable of benefits or merit of any received piece of information. Structures are formed as construct that is found on lexis and grammar of a given language. Formally, analyzing any language as a tool for building structures requires definitions, atomic elements and operations/rules for building structures. The more complex a structure becomes, the more semantically refined language expressiveness is of the language used for its development.

Upon investigating a possibility of automation of structure building using atomic elements, it is necessary to consider a requirement of formal definition of structure-specific language operations. If the same structure can be build in a variety of possibilities it means that the language operations can lead to vagueness in interpretation of the given structure. The degree of structure complexity and acceptable differences of interpretations indicate language expressiveness degree.

Limiting the quantity of atomic elements and operations/rules of structure building guarantees maintaining higher degree of unambiguity of understanding. Semantic expressiveness, clarity and unambiguity in interpretation of essence of given structure are contradictory and depend on the degree of language formalization.

Surrounding reality is defined by varied and infinitively large collection of information, or in other words stimuli received by an individual. Cognitive science assumes natural ability of human mind to categorize information absorbed from the exterior environment and from human memory into the form of concepts. According to [26], a concept is considered as a structure of knowledge representation for the mind that contains description of significant features of given category (class). Furthermore, knowledge contained by the mind is stored in form of concepts. That phenomenon can be observed when we relate to abstract knowledge which often does not have a physical representation and forms a collection of concepts pertinent to some idea. Tutoring with conceptual knowledge is based on maintaining communication among participants of that process preserving the conceptual aspect of that process. In addition, it is possible to leverage metaphors and other pedagogical methods. However, it is important to preserve the requirement of referring to the same concepts upon communicating among minds of both types of the parties involved in the communication process.

Knowledge representation model includes: definition of semantic elements, operations performed on them and a knowledge representation language specific to those needs. The distance learning environment does not provide an inclusion of natural language to its full extent as a tool for knowledge relay. That situation contributes to forming a premise for constructing a specialized language. With accordance with the aforementioned arguments, the model of knowledge representation and relay in the distance learning environment should guarantee and facilitate the following:

- correspondence with natural language and languages of representation of knowledge originating from varying resources
- maintain a sufficient level of unambiguity involved in the process of interpretation
- maintain sufficient semantic expressiveness enabling creation of complex structures
- provide an ability for structure contents correction, depending on a context
- provide an ability for creating complex structures using already available simple structures
- provide a mechanism of joining of distinct structures and forming a semantic network of specified and limited semantic linkage.
- provide means for connecting a single objective domain knowledge model with corresponding method of instruction

The fundamental entities of knowledge representation language is a concept - a logical unit that is an atomic knowledge element. Every concept represents certain class of objects with similar attributes.

Fully functional language also includes the mechanisms for performing operations on concepts. It includes the operations of extending semantics, i.e., adding a new attribute, or adding a new object to the knowledge contents and the links between concepts, i.e., semantics of links. Using language defined as such, for every domain a lattice (graph) is formed in which the vertexes represent a concept of a given domain.

5.2.3 Fundamental and Operational Knowledge

Academic teaching requires effective distinction between fundamental and operational knowledge. Specification of areas dedicated to the given type of knowledge allows to apply an appropriate strategy and method of teaching. The requirement of fulfilling that distinction is consequence of adapting the learning-teaching process that is different for fundamental and operational knowledge. However, difficulties will likely emerge upon necessity of effective settling both types of knowledge since each of the academic courses contains proportionally components of fundamental and operational knowledge. Academic teaching, both on university and polytechnic levels of education, the component of fundamental knowledge plays a dominant role.

Knowledge researched as an object and purpose of teaching can be defined as two fundamental kinds: fundamental-theoretical and operational (that problem has been thoroughly investigated in [21]). The kind of fundamental knowledge represents conceptual thinking leading to establishing new paradigms, new problems and problem statements, conduct protocols, etc. The type of operational knowledge is necessary for realization of scenarios and algorithms of maintaining operations. Situations that a human individual encounters in one's day-to-day operations involve both kinds of knowledge in action in varying proportions, depending on the analyzing operations degree of complexity. Existing methods of knowledge representation and computer technologies offer opportunities for constructing a sufficiently effective teaching system of operational knowledge (e.g. computer simulators, expert systems and others}. It is feasible because it is possible to adequately join quantitative assessments of skills acquired by students (speed, precision) with corresponding theoretical knowledge base. Teaching using fundamental knowledge is still strained because there is a discernible lack of effective quantitative methods suitable for assessment of an increase of student's fundamental knowledge.

Fundamental knowledge presented by many researchers including [30] can be divided into tacit knowledge (the one that human individuals are not constantly aware of) and explicit knowledge. The tacit type, as described by [4] is a subtle, personal knowledge engaging one's own system of beliefs, understanding of the outer world and its valuation. Informal knowledge (tacit knowledge) is associated with one's own experience providing means for executing a variety of tasks. Formal knowledge (explicit knowledge) is expressed through language and relayed as information associated with procedures, data bases and patents. Genuine value of tacit knowledge is attributed with its complex nature. However, as pointed out by [35] upon analyzing the cost of knowledge transfer, the more formalized knowledge is, the more economical the transfer becomes. It is due its unambiguous interpretation, clearer structure and contains fewer unclear elements.

5.3 Ontology Serving as Knowledge Model

Ontological approach towards knowledge is currently considered as one of key elements in the actively developer idea of information-based society. The main premise for formulating that thesis is the creation and dynamic development of the concept of Semantic Web [2] and also business-class systems that utilize the apparatus of ontology ([10], [27]). Semantic Web enables the computers equipped with mechanisms pertaining to the domain of artificial intelligence to assume an analysis of contents of any given web page to process new knowledge.

Upon investigating the problem of knowledge acquisition and its proper organization it has been acknowledged that ontologies play a fundamental role in development of a model providing real-world knowledge representation. Computer ontologies have become an apparatus guaranteeing mutual agreement on the concepts included in the model not only among experts from different field of study, but among users operating with the same methods and tools. In that sense, the ontology provides means of interpretation for some of domains and also allows to determine a yardstick for communication among human individuals and other application systems. Those possibilities are extensively developed and used and that has been shown by emergence of ontology formalization languages and the tools supporting its development, unification and processing.

The expansion related to the range of applications of ontology has spurred emergence of practical solutions such as WordNet and EuroWordNet [12] primarily for the needs of linguistics; UMLS for medicine [5]; EngMath for engineering applications [13]; OTA for tourism [39] and Enterprise Ontology [37] for pursuing formal descriptions of business logic of enterprises.

5.3.1 Definition of Ontology

A discussion on the meaning of term ontology should be initiated by presenting its etymology. Its origins comes from Greek and the term itself is a combination of two terms: ontos - existence, entity, live being, or functioning and live organism and logos - mind, word, science created from the verb lego meaning to reckon, calculate, reason, speak, collect and to merge.

Although the concept was presented in the vocabulary of philosophy as late as in the XVIIth century in form of a commentary to the term abstraction, the precursor of research activities involving ontologies was Parmenides. Around V-IV century BC he acknowledged independent character of the essence when confronted with human senses. When it comes to identifying the first creator of ontology, the literature points to Aristotle who devised and presented a system for categorization of virtually everything that can be asserted about the world. The term "ontology" was popularized only in XIII-th century by Christian Wolff in his philosophical dissertation entitled 'Philosophia prima, sive Ontologia'.

In the last decade the term has become to be applied among knowledge engineering professionals. One of the earliest definitions of ontology was proposed by Neches and stated as "ontology defines both the elemental concepts and relations forming a dictionary of the given thematic area and the rules of merging the concepts and for dictionary expansion" [13]. In that sense, not only does the ontology entail all of the concepts, but also knowledge that can be inferred from that ontology.

One of the most frequently quoted definitions was proposed by Gruber, according to which, the ontology is a specification of conceptualization [14]. That definition has laid the groundwork for ensuing discussion concerned with the meaning of ontology. Further expansions of the definition was offered by Borst that defined the ontology as a formal and unambiguous specification of a shared conceptualization [12]. That interpretation was also used and defined by Studer and others [34]. According to the latter, the term of conceptualization refers an abstract model of determined events in the real world through carrying out an identification of appropriate concepts characterizing those events, whereas the aspect of unambiguity results from the fact the concepts used in that conceptualization and limitations on their use are precisely defined. The term "formal" in that sense refers to the requirement of forming a machine-readable (computer) ontology. The shared aspect of the term relates to the fact of accepting by recipients the concepts that are used for the purpose of maintaining the ontology.

As development of understanding of the term proceeds, it is reasonable to consider ontology as a science dealing with 'kinds and structures of objects, attributes, events, processes and relations i every aspect of surrounding reality [31]. According to Fensel, the principal task of the ontology is providing means for developing a model of certain knowledge domain [9]. The ontology is supposed to be a viable tool for conducting domain knowledge conceptualization which itself offers a possibility for modeling the essence of the domain-specific concepts and relations existing among them (see figure 5.1.). The ontology should provide mechanisms both for extracting a class of objects required to conduct exhaustive conceptualization of given knowledge domain and for inferring on the occurring processes. That particular approach was expanded by Heylighen acknowledging the ontology as an outcome of abstract analysis of conceptual system that can be also represented as a graph. The vertexes of the graph represent fundamental concept identified during the analysis and the arcs represent the relations between the concepts [12]. The final outcome of that form of analysis performed using the aforementioned approach is an ontology.

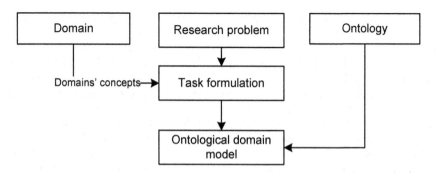

Fig. 5.1 Ontology and related ontological model

Domain ontology is an outcome of representing a relatively limited real-world sample and is dependant on the approach presented herein. That feature contrary to other categories of ontology, such as the top-level ontologies, or representational ontologies that represent, respectively, generally available conceptual structures and meta-structures [25]. The main motivating force for development of domain ontologies is the ability to share and reuse of domain-specific knowledge among different software and in different real-world applications. In the work of [15] a strong refutation of the objection claiming impossibility of developing a domain ontology that could face the purpose of serving a variety of different application scenarios.

The concept of ontology is premised on widely acceptable findings binding structure and behavior of real objects. The premises serving as a basis for proper understanding of the concept have been presented in the works of [6] and are the following:

-   objects exist in the surrounding reality
-   the objects have features and attributes that can assume different values, for example as in the sequence object - attribute - value
-   objects are associated with each other through a variety of relations
-   both the relations and the attributes can be subject to change as time progresses
-   events may occur immediately and in different time frames
-   there are processes occurring place constantly involving objects
-   the world and its objects may have different states
-   the events can cause or lead to other events or object states
-   the objects may be composed of parts

Typical causes for developing the ontology have been formulated in [29]. Those are:

-   sharing of mutual understanding of information structure among individuals or internet agents
-   enabling domain knowledge reuse
-   enabling creation of assumptions consciously about the given domain
-   enabling separation of domain knowledge from operational knowledge entailed by that domain
-   enabling conducting analysis of domain knowledge

Analysis of the causes clearly shows a high degree of conformity with the domain of distance learning, especially upon consideration of the Learning Object approach. During the process of developing a instance of Learning Object the key element is successful development of domain knowledge model. That task is concerned with identification and association of concepts and relations that exist in the given domain. Ontological analysis provides statement of the domain knowledge structure and enables to determine a method of knowledge encoding sufficient for maintaining both the concepts and the relations [6].

5.3.2 Classification of Ontologies

The very first ontologies created for the needs of knowledge engineering started to become developed in 1980s. Intense development scope resulted in development of a number of ontology categories. Aiming to formulate an appropriate categorization, Gomez-Perez et al. [12] has proposed the following list of ontology types (figure 5.2): top-level ontology, domain ontology, task ontology and application-driven ontology. Each of the types can be characterized with varying degrees of generality. The top-level ontology captures knowledge about the world providing elemental concepts dealing with time, space, event and state. Those terms are common and universal and can be used to develop specific domain ontologies. The domain ontologies present knowledge that is characteristic to a given domain, such as medicine, pharmacy, law or music. The concepts used to carry out the conceptualization are specialized products of the top-level ontology. The task ontology forms a dictionary that is unique to specific task or activity, such as conducting diagnostic testing, or scheduling by applying specialized the terms derived form top-level ontologies. In contrast with the domain ontology, in this

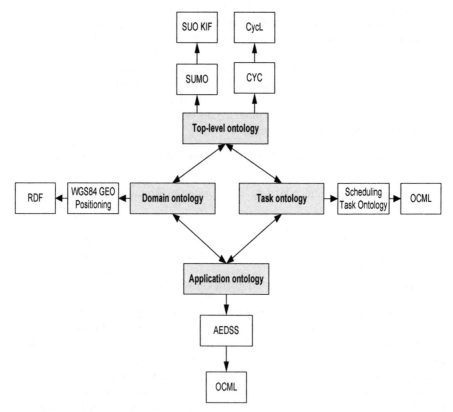

Fig. 5.2 Types and instances of ontologies and the language used for carrying out ontology formalization (based on [19])

case to solve the given problem effectively an ontology of that type may combine concepts from different domains. The application-driven ontology contains definitions of concepts that are required for conceptualization of knowledge used in a variety of application scenarios. It forms and specializes the dictionary of the domain and task ontologies in the context of a particular application.

The outcome of categorization of ontologies are two approaches toward ontology expansion that have their origins in the body of philosophical sciences: the top-down approach and the bottom-up approach [12]. The former requires expanding an ontology through, in first place, identification of a general classification of entities that is common for all of the domains.

However, a limited likelihood exists for finding a sufficient top-level ontology that would satisfy all the assumptions about all of contexts, application scenarios and domains. On the other hand, that particular kind of solution would not only allow to give a proper shape to top-level ontology, but would also allow to infer using detailed conceptualizations of domains. The inference process with accordance to the bottom-up approach recommends to first start with expanding the domain ontologies and after that proceed with forming conclusions concerning a more general ontology. A lack of uniform and commonly acceptable method for development of ontology for specific domains leads o potential risk of inability to form a more general ontology through generalization of the concepts.

5.3.3 Ontology Engineering

All of the problems occurring during development and management of conceptual domain model can be generalized into a single category of ontology engineering (figure 5.3.). The term has been defined by [12] as a collection of activities related to the process of development of ontology, ontology life cycle, methods and methodology of developing ontologies and languages supporting all of the aforementioned processes. Ontology engineering offers a mechanism to form collections of significant concepts of given domain and to form them into a structure using statistical combination and linguistic data. The environment of ontology engineering is based on the object-oriented paradigm. A developed object-oriented model can be analyzed and modified with operation with a corresponding concept representation and the corresponding relations.

The approach used in for the purposes of ontology engineering significantly differs from the approach used upon developing traditional data models such as those found in the works of [36]. Data model represents a structure that needs to satisfy the requirement of integrity for a given scenario and context the model has been developed for. As it has been argued by [33], data model is not aimed to a priori satisfy the requirement of being shareable for different computer applications. Data models are targeted for particular real-life applications. They are often products of discussion between a model engineer and its recipients (users). When a need arises, further expansion of the model can be facilitated through a properly designed modification, for example using XML files or specific database structures.

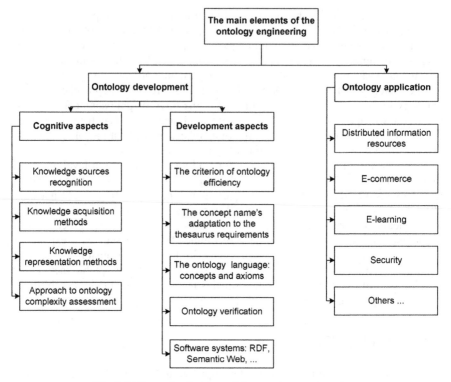

Fig. 5.3. The main elements of the ontology engineering

Similarly as for the development of the Learning Object, the greatest challenge for the ontology engineering is satisfying the requirement of universality. Data model is oriented at particular applications scenario and implementation and its quality is evaluated on the basis of the most effective match of the model to the real-life application. Contrary to the data model, the ontology should remain as much general and application-independent. At the same time, the ontology should also provide conceptual contents of the given domain. As it has been discussed by [3], the ontology should maintain the highest degree of generality for the needs of relay of detailed and specialized knowledge. That paradox is challenging to resolve since the boundaries of the generality need to be identified (all of the considerations relate to the area of identified by boundaries of a given paradigm). The quality indicator of the ontology in question, greatly influenced by the degree of generality, can be the ability to provide genuine domain knowledge in a way that would make it accessible for different computer and practical applications in the given domain.

5.3.4 *Ontological Knowledge Specification*

The Conceptual domain scheme using in distance learning for Learning Object development is a structure that contains domain expert knowledge in a form of ontology. The expert defines the elements of ontology using abstraction operations. Each of the concepts is an abstraction with a determined depth of a domain concept. Operation of aggregation and generalization impacts the determination of relation among domain concepts and defining a dimension for every concept by association with a context.

Forming the conceptual domain scheme in a form of ontology is based on the process of knowledge domain conceptualization using classes, instances, relations, functions and axioms specified for the given domain [38]. Usually, that process necessitates development of a conceptual hierarchy and assignment of their attributes. The conceptualization process in particular requires development of concepts $\{c_1,...,c_i,...,c_n\}$ that specify the ontology. In the given set the c_i element is, respectively, a representation of a class, instance, relation, function, and a representation of an axiom.

According to [38] and [16], ontology is defined as the following tuple: $W = <CD, ID, RD, FD, AD>$. Identification of the listed elements indicates, as in [18]:

- CD – set of class definitions containing concepts used for describing of real world;
- ID – set of instances definitions, i.e., set of real-world instances of the objects identified in CD;
- RD – set of relations defined with respect to the set CD. The relations are divided into the relations that belong to taxonomical structures of the domain (structural relations), and the relations that are otherwise (non-structural relations);
- FD – set of functions definitions that are executed in the set of the concepts yielding to new concepts. The set can be considered as a set of inference rules;
- AD – set of axioms definitions.

A common practice exemplified by the works of [18] is limiting the definition to the following tuple: $W = <CD, ID, RD>$. The FD and AD sets are neglected and the ontology defined as such is called a lightweight ontology. The lighwieght ontology knowledge model is expressed by means of the sets of concepts, instances and relations that interconnect them. Conducting modeling of the conceptual domain scheme using lightweight ontology leads to certain simplifications that lead to improving the clarity of the ontology. In that case, the domain expert is not obliged to conducting search of artificial phenomenons that constitute the set of AD. The lightweight ontology somewhat reflects traditional perception of knowledge (namely through the sequence of concept and relation) that is exemplified in the semantic networks knowledge representation method.

5.3.5 The Degree of Ontology Generality (Example)

The categorization of ontologies enables to reproduce a mode of thinking and after that to state a problem and propose its solution. It should considered that not only an origin of the problem is subject to analysis, but also (and more importantly) the domain that is pending a problem resolution. While using ontology as a means for addressing a certain problem it is necessary to proceed through subsequent degrees of generality in order to initiate interpretation of given domain knowledge using terms acceptable for that domain, or conforming to the chosen tools of problem resolution.

Let us investigate that approach using an example (see figure 5.4.). The first degree of generality defines an overall characteristic of a problem and enables the problem to be stated. From the process modeling perspective, the next categories of the ontology need to define a domain that the processes relate to (manufacturing systems). Since we may deal with a variety of processes forming research objects, the specifics of the task are formed (stochastic processes). Only after forming a whole element it offers a possibility of establishing an approach towards resolving the problem (simulation modeling).

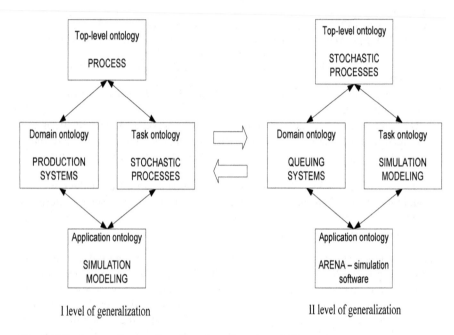

Fig. 5.4 The degree of generality of ontology (based on [19])

The solution requires conducting an analysis because a way of solving the problem needs to be determined. Therefore, proceeding to the second degree of generality is required. At that stage, all the stochastic processes become the main category of research. Those processes may be formally described using the domain queuing systems theory and as such can become a subject of further analysis. In that context, task statement is required as a starting point for proceeding further to another categorization level. Definition of a task in the domain of simulation modeling serving as ontology indicates a requirement for developing a specific simulation model. Understanding of type of simulation to be run can lead us to finding the method of finding the solution, which in discussed case proves to be the Arena simulation software.

The defined ontologies reflect the state of knowledge and indicate its subsequent levels of analysis and specialization. However, forming and presenting result of one to domain and presenting then using terms of the other is a significantly different task. Upon formulating the task definition that aims at interpreting the simulation results and transposing them onto the Kendall's notation of queuing systems, it appears that the nomenclature of functions in the Arena software diverges from typical names of concepts both in the queuing systems theory and the Kendall's notation. That situation demands performing identification of direct counterparts between two conceptual models. To deal with that kind of task, a dedicated process of concept mapping enabling to establish relations between concepts in two different domains has been proposed in [21]. Figure 5.5. illustrates the procedure.

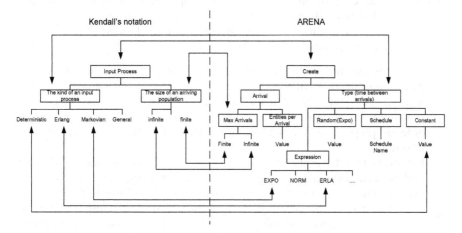

Fig. 5.5 Mapping of the concept "input stream" between domains of Kendall's notation and Arena software (based on [21])

The process of concept mapping between two distinct ontologies is a very complicated task due to the following factors [8]:

- different applications of the concepts in the given ontologies
- different semantical range of the concept
- different semantics resulting from applications in different natural languages, or more likely due to different classification of the concept in the given ontology
- different semantical relations

Therefore, upon performing concept mapping between the ontologies, the semantic range of given taxonomies is necessitates an analysis.

5.3.6 Extended Ontological Model

Upon considering the possibility of creating an ontological model for distance learning that preserves reusability, one has to pay special attention to the necessity of presenting formal definitions of both adequate operations and the model (structure). In distance learning we deal with such situation when using a structure we try to reflect a domain knowledge the same with the corresponding pedagogical and cognitive requirements. To fulfill these requirements the extended ontological model was developed and described in detail in [20].

When domain knowledge is the object of structuralization, domain concepts that can be sorted in different ways become the main model components. The purpose of the domain knowledge structure is to ensure conformity of the knowledge representation model with the student's cognitive profile and learning goals. This is exactly the kind of mission a teacher carries on during direct interaction with a student. Another – just as important – task results from the following sequence: providing the possibility of adapting the knowledge structure to the personal learning situation when the teacher is not available.

The arguments above lead us to consideration of the following conditions that influence the extended ontological model in distance learning [20]:

- Upon creating the ontological model, it is vital to differentiate knowledge types: fundamental (theoretical) from procedural (operational), which together constitute the domain knowledge. Analysis of any individual situation is based on simultaneous usage of both kinds of knowledge and in different proportions – depending on the level of problem's complexity and its origin.

- Following David Ausubel, the author of the concept of meaningful learning [28], we can claim that learning is concerned with assimilating new concepts to the cognitive structures (conceptual maps) already existing in the student's mind. Contrary to rote learning, learning through understanding (meaningful learning) use discovery learning process [17], where all concepts' attributes are individually identified by the student. In order for that to happen the following requirements must be met:

  o learning material needs to be conceptually refined, language and examples should be adapted to knowledge already possessed by the student,
  o the student has to be able to recognize and understand the linkage between new material and his/her knowledge,
  o the student has to make effort and choose meaningful learning by himself/herself.

- While designing a distance learning system based on knowledge it is especially necessary to pay attention to the following sequence: information – knowledge – competency. Competences and qualifications are one of the basic mechanisms of assessing knowledge assimilated by the student [24]. Competencies are always associated with a certain person and represent his/her knowledge and ability to use that knowledge in a certain domain scenario. On the other hand, qualifications reflect the competences basing on some assessment system. The qualification system defines a hierarchy of qualifications, where the bottom qualification includes noticeably smaller amount of competencies than the advanced one. In addition, every qualification is related to a corresponding set of competencies.

- Development of a system which would ensure human-computer interaction requires a detailed analysis of the given problem's cognitive aspects. Upon analyzing an ontological knowledge system pending development (special attention paid to the distance learning) it is important to consider the following factors:
  - o the way different types of knowledge are stored in our mind,
  - o mind-related organization of the information – knowledge processing sequence,
  - o human perceptual constraints,
  - o cognitive learning styles.

Extending of the ontological model, as suggested by the author, deals with adaptation of a typical ontology to the requirements in distance learning and also considering the demand for reusability. The intention of the authors is enabling an extension of the ontological model preserving learning goals. With regards to current state of knowledge it is not yet possible to create an universal ontology that could be used in different contexts. Therefore, it is necessary to create a methodology for dedicated model management. We assume a typical scenario of working with the knowledge model. Its first step is executed by the teacher deciding on the scope of knowledge. The choice of that scope can be narrowed to defining the set of concepts which will become a foundation for creating a didactic material. In the second step, the semantic depth - a scope of competences to be introduced to the student – is defined for each concept.

   The ontology structure of the extended ontological model is composed of:

- the way of describing concepts structure.
- the agreed semantic relations between concepts.

The relations have been divided into taxonomic and non-taxonomic types and also to a set of axioms of the modeled domain. A thesaurus defines the vocabulary that can be used while defining concepts and relations and also references for concepts and relations. Taking all of the presented assumptions and assertions we may formally introduce the concept of ontological model (OM):

$$OM = \{S, T\},$$

where: S – ontology structure, T – domain description thesaurus,

$$S = \{Sc, R, T, A\},$$

where: Sc – concept structure, R – relation between concepts, T – taxonomy, A – set of domain axioms.

$$T = \{Tp, Tr, Fp, Fr\},$$

where: Tp – thesaurus for the set of concepts, Tr – thesaurus for the set of relations, Fp – references for concepts, Fr – references for relations.

   When developing knowledge models presented to the student in distance learning systems, it is necessary to preserve the goal of learning concerned with the required competences and the student's basic knowledge. The level of knowledge capacity corresponds to the subset of objects counted to the class which was nominated by the name of the concept. The level of knowledge depth corresponds to the set of characteristics of each object recognized as a member of a certain class, with consideration of constraints forced by a specified goal of learning and the required competencies.

In literature we may come across many definitions of a concept. The main characteristic of a concept is usually its mental representation structure created through abstraction and generalization. [36] and [40] have proposed an interpretation of a concept as a nomination of a certain class of objects which share similar features. In our research, the following definition of a concept will be used [22]: a concept is a nomination of classes of objects, phenomena, or abstract category and for each of them the common features are specified in such way that distinguishing each class does not cause any problem.

The approach to extended ontological model development is based on using semantic operations such as PART_OF, IS_A, KIND -OF (see figure 5.6.). Both aggregation (PART_OF), generalization (IS_A) and specialization (KIND_OF) are semantic operations, which can be considered as results of method of abstraction creation. In that context, a concept can be seen as an abstraction [11] and that should help understanding of a complex object through decomposing it into less complicated components. Thanks to the PART_OF relation it is possible to describe the set of characteristics sufficient to recognize a certain abstract object as a member of the class under investigation. The IS_A operation claims that a specific object with the given values has been included as a member of the same class. Lastly, the KIND_OF operation means that specific objects listed by name have been included in the considered class.

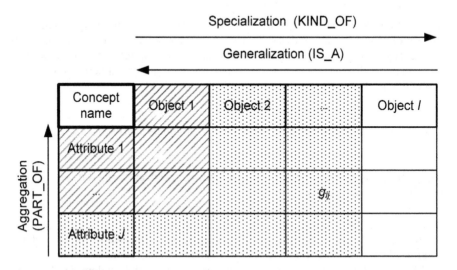

Fig. 5.6 Matrix-based structure of a concept (source [20])

Formally the content, capacity and depth of knowledge can be described in the following way as a matrix concept structure:

$$G = \left\| g_{ij} \right\|, i = \overline{0, i^*}, j = \overline{0, j^*}, \text{ where:}$$

$$g_{ij} = \begin{cases} \hat{G}, i = 0, j = 0, \text{concept's name} \\ O_i, i = \overline{1, i^*}, j = 0 - \text{names of objects belong to class } \hat{G}, \\ W_i, i = 0, j = \overline{1, j^*} - \text{names of common attributes belong to class } \hat{G}, \\ g_{ij}, i = \overline{1, i^*}, j = \overline{1, j^*} - \text{value of feature } j \text{ for object } i, \end{cases}$$

then: the tuple $< \hat{G}, W_1, W_2, \ldots, W_{j^*} >$ describes the content of knowledge corresponding to the term of concept $\hat{G}$. Tuple $< \hat{G}, O_1, O_2, \ldots, O_{i^*} >$ describes the capacity of knowledge corresponding to the term of concept $\hat{G}$. The set of values of all features of all objects $\{ g_{ij} \}$ is a description of the depth of knowledge corresponding to the term of concept $\hat{G}$. Adding new elements to set $I = \overline{1, i^*}$ while maintaining the content of the concept means broadening the examined class of objects. Adding or removing elements from set $J = \overline{1, j^*}$ means changing the content of the concept. Intersection $\delta = I \cap J \neq 0$ is the measure of acceptable tolerance for different forms of the concept, which correspond to the border of the domain being examined. When $\delta = I \cap J = 0$ we deal with a situation in which the same word in different knowledge domains refers to a different thing or phenomenon. The adequate extended ontological model creation algorithm has been discussed in detail in [41].

Development of an ontological domain model in an educational scenario requires analyzing a specific curriculum and learning goal, which constrain both the capacity and the semantic depth of concepts used in didactic materials. In that context, the presented definition of ontology that utilizes matrix structure of a concept leads to a two-tier layout of the ontological domain model [20]. The first level is a network of concepts (similar to a semantic network) and the second level defines the depth and capacity of knowledge contained in each concept. The rules for creating the two-level arrangement can be used many times in reference to the same originally defined ontological model. That gives the possibility of developing a multi-level ontological model. Using the proposed approach enables adjusting the ontological domain model to specific educational goals. The ontological domain model extended in such way will allow for a significant level of automation of processing the model into a modular structure of didactic materials that are dedicated to be used during the learning process.

As an example of the proposed matrix-based structure of a concept, Table 5.1 illustrates a problem of capacity and depth of a concept of the queuing system domain utilizing the Kendall's notation. The matrix presents the details of the concept of queuing systems that are an organized form of certain stochastic processes oriented towards arrival of incoming assignments/jobs/tasks. Basic operational parameters of those systems are: arrival pattern, kind of servicing, number of servers, system capacity, population and discipline of servicing.

Various systems of different element combinations (elements such as servers, buffers, users, and others) can be classified into the class of queuing systems:

-   M/M/1 – Markovian single-server system of infinite capacity, population and the FCFS (first came first served) discipline of servicing
-   M/M/m – Markovian multi-server system of infinite capacity, population and the FCFS (first came first served) discipline of servicing
-   others according to the Kendall's notation

Every column of the matrix contains a specification of the researched stochastic system. The system can be interpreted as a queuing system of certain structure and organization. The presented specification is sufficient for the requirements of a proper mathematical apparatus [42].

Table 5.1 Example of matrix concept's structure of queuing system concept (based on [20])

Queuing Systems	M/M/1	M/M/m	M/M/m/m	M/M/1/S	M/D/1/C	G/G/∞/prt	...
Arrival pattern	Markovian (Poisson) arriving process = M(P)	M(P)	M(P)	M(P)	M(P)	General	...
Kind of Servicing	Markovian (Exponential) servicing process = M(E)	M(E)	M(E)	M(E)	Deterministic	General	...
Number of servers	1	m	m	1	1	∞	...
System capacity	∞	∞	m	Limited by S	∞	∞	...
Population	∞	∞	∞	∞	Limited by C	∞	...
Discipline of servicing	First come first serve (FCFS)	FCFS	FCFS	FCFS	FCFS	Priority	...

According to the table 5.1. the tuple (concept's designation with a set of associated attributes) is an intensional definition. It reflects the contents of the concept of queuing systems. The collection of column names such as M/M/1, M/M/m, M/M/m/m,... G/G/∞/prt,..., define the concept's semantical capacity. The set containing values of all elements stored in the matrix constitute the semantical depth of the given concept. Definition of the concept presented in Table 5.1. can be utilized upon creating an ontological model. The proposed conception of introducing concepts' definitions should be utilized only as one of cognitive rules of knowledge module development for the Learning Object.

5.4 Conclusions

The proposed methods of knowledge and ontology engineering addressing the needs of distance learning alter the methodology of learning materials development. The actors participating in the process are subject of that change as well as are their roles that simply become overdue. All of the indicated changes require conducting an analysis in the light of requirements and pedagogical determinants of the teaching-learning process. Some relations and procedures cannot be ignored for example during an attempt to maximize automation of the process.

Both procedural and fundamental knowledge require separate learning environments. Instruction on the area of given subjects necessitates consolidation of those two types of knowledge. That requires an ability to develop unified and corresponding platform for both types of knowledge and also development of its effective use in practice.

References

1. Anderson, J.R.: Cognitive Psychology and Its Implications, 5th edn. Worth Publishing, New York (2000)
2. Berners-Lee, T., Hendler, J., Lassila, O.: The Semantic Web: A new form of Web content that is meaningful to computers will unleash a revolution of new possibilities. Scientific American (May 17, 2001)
3. Borst, P., Akkermans, H., Top, J.: Engineering ontologies. International Journal of Human-Computer Studies 46, 365–406 (1997)
4. Carvalho, L.C., Rodrigues, M.E., Paret, B.D.: Tacit and Formal Knowledge and Learning in Small Business: An Exploratory Study on the Perceptions of Successful Businessmen. In: Proceedings of 4th International Conference on Technology Policy and Innovation, Curitiba, Brasil, August 28-31, pp. 28–31 (2000)
5. Ceusters, W., Smith, B.: Ontology and Medical Terminology: Why Description Logics are not enough. In: Proceedings of the Conference Towards an Electronic Patient Record, TEPR (2003)
6. Chandrasekaran, B., Josephson, J., Benjamins, V.: What Are Ontologies, and Why Do We Need Them? IEEE Intelligent Systems 14(1), 20–26 (1999)
7. Codd, E.F.: Further normalization of the data base relational model. In: Rustin, R. (ed.) 6th Courant Computer Science Symposium: Data Base Systems. Prentice-Hall, New York (1872)

8. Doerr, M.: Semantic Problems of Thesaurus Mapping. Journal of Digital Information 1(8) (2001)
9. Fensel, D.: Ontologies: Silver Bullet for knowledge Management and Electronic Commerce. Springer, Berlin (2001)
10. Fensel, D., Horrocks, I., Harmelen, F., McGuinness, D.L., Patel-Schneider, P.F.: OIL: An Ontology Infrastructure for the Semantic Web. IEEE Intelligent Systems 16(2), 38–44 (2001)
11. Goldstein, R.C., Storey, V.C.: Data abstractions: Why and how? Data & Knowledge Engineering 29(3), 293–311 (1999)
12. Gomez-Perez, A., Corcho, O., Fernandez-Lopez, M.: Ontological Engineering: with examples from the areas of Knowledge Management, e-Commerce and the Semantic Web. Springer, London (2004)
13. Gruber, T., Olsen, R.: An Ontology for Engineering Mathematics. Morgan Kaufmann, Gustav Stresemann Institut, Bonn (1994)
14. Gruber, T.R.: A translation approach to portable ontologies. Knowledge Acquisition 5(2), 199–220 (1993)
15. Guarino, N.: Understanding, building and using ontologies. International Journal of Human-Computer Studies 46, 293–310 (1997)
16. Heflin, J., Hendler, J.: Dynamic Ontologies on the Web. In: Proceedings of American Association for Artificial Intelligence Conference AAAI 2000, pp. 443–449. AAAI Press, Menlo Park (2000)
17. de Jong, T.: Scientific Discovery Learning with Computer Simulations of Conceptual Domains. Review of Educational Research, 179–201 (1998)
18. Koprowska, M., Juszczyszyn, K.: The ontology merging in the semantic environment. In: Proc. V National Conf. Inżynieria Wiedzy i Systemy Ekspertowe, vol. 2, pp. 294–301. Wrocław, Poland (2003) (in Polish)
19. Kusztina, E., Różewski, P., Ciszczyk, M., Sikora, K.: The ontology considered as a tool for domain knoeledge description. Metody Informatyki Stosowanej 12(2), 73–88 (2007) (in Polish)
20. Kushtina, E., Różewski, P., Zaikin, O.: Extended ontological model for distance learning purpose. In: Reimer, U., Karagiannis, D. (eds.) PAKM 2006. LNCS (LNAI), vol. 4333, pp. 155–165. Springer, Heidelberg (2006)
21. Kusztina, E.: Conception of Open Information System for Distance Learning, Szczecin University of Technology. Faculty of Computer Science and Information Systems (2006) (in Polish)
22. Kusztina, E., Zaikin, O., Różewski, P., Tadeusiewicz, T.: Conceptual model of theoretical knowledge representation for distance learning. In: proceedings of the 9th Conference of European University Information Systems (EUNIS 2003), Amsterdam, pp. 239–243 (2003)
23. Lacey, A.R.: A Dictionary Of Philosophy. Routledge, Chapman&Hall (1996)
24. Lahti, R.K.: Identifying and integrating individual level and organizational level core competencies. Journal of Business and Psychology 14(1), 59–75 (1999)
25. Maedche, A., Staab, S.: Discovering Conceptual Relations from Text. In: Technical Report 399, Institute AIFB, Institute AIFB, Karlsruhe University, 76128 Karlsruhe, Germany, vol. 400, pp. 321–325 (2000)
26. Maruszewski, T.: Psychology of recognition. GWP, Poland (2002) (in Polish)
27. McGuinness, D.L.: Ontologies and Online Commerce. IEEE Intelligent Systems 16(1), 8–14 (2001)

28. Novak, J.D.: Learning, Creating, and Using Knowledge: Concept Maps As Facilitative Tools in Schools and Corporations. Lawrence Erlbaum Associates, Mahwah (1998)
29. Noy, N.F., McGuinness, D.L.: Ontology Development 101: A Guide to Creating Your First Ontology. Knowledge Systems Laboratory Stanford University (2001)
30. Polanyi, M.: The Tacit Dimension. Routledge & Kegan Paul, London (1983)
31. Smith, B.: Ontology. In: Floridi, L. (ed.) Blackwell Guide to the Philosophy of Computing and Information, pp. 155–166. Blackwell, Oxford (2003)
32. Smith, J.M., Smith, D.C.P.: Data base Abstractions: Aggregation and Generalization. ACM Transactions on Database Systems 2(2), 105–133 (1977)
33. Spyns, P., Meersman, R., Jarrar, M.: Data modelling versus ontology engineering. ACM SIGMOD Record 31(4), 12–17 (2002)
34. Studer, R., Benjamins, V.R., Fensel, D.: Knowledge Engineering: Principles and Methods. IEEE Transactions on Data and Knowledge Engineering 25(1-2), 161–197 (1998)
35. Teece, D.J.: Capturing Value from Knowledge Assets: The New Economy, Markets for Know-How and Intangible Assets. California Management Review 40(3), 55–79 (1998)
36. Tsichritzis, D.C., Lochovsky, F.H.: Data models. Prentice Hall, Englewood Cliffs (1982)
37. Uschold, M., King, M., Moralee, S., Zorgios, Y.: The Enterprise Ontology. The Knowledge Engineering Review 13 (1998)
38. Visser, P.R.S., Jones, D.M., Bench-Capon, T.J.M., Shave, M.J.R.: An Analysis of Ontology Mismatches; Heterogeneity versus Interoperability. In: Working Notes of the Spring Symposium on Ontological Engineering AAAI 1997, pp. 164–172. Stanford University, Stanford (1998)
39. Vukmirović, M., Szymczak, M., Ganzha, M., Paprzycki, M.: Utilizing Ontologies in an Agentbased Airline Ticket Auctioning System. In: Proceedings of the 28th ITI Conference, pp. 385–390. IEEE Computer Society Press, Croatia (2006)
40. Wong, H.K.T., Mylopoulos, J.: Two Views of Data Semantics A Survey of Data Models. Artificial Intelligence and Database Management. INFOR 15(3), 344–382 (1977)
41. Zaikin, O., Kusztina, E., Różewski, P.: Model and algorithm of the conceptual scheme formation for knowledge domain in distance learning. European Journal of Operational Research 175(3), 1379–1399 (2006)
42. Zaikin, O.: Queuing Modelling Of Supply Chain In Intelligent Production. Wydawnictwo Informa, Szczecin (2002)

Chapter 6
Learning Object Methodology

6.1 Introduction

Etymology of the term "repository" defines that object as a place intended for storage of records and official materials preserving the ability to use them. Present resources significantly extend the nature of functioning of repositories [37]. Currently, they are considered as a place for storing collections of digital documents and sharing them in a network for a selected group of people or with unlimited access with an aid of appropriate user interface [21,20,33].

The main purpose of the repository is to ensure for each user is provided with a possibility to read, copy, distribute, print, search or link full texts of scientific articles and other materials and documents placed in the repositories [20]. That view of the object is strongly supported by the members of the Open Archives Initiative (OAI) [5].

The existing repositories can be used for the purpose of learning process as a source of domain outlook, research means of development or as a source of didactic materials. By using the repository, elements of domain knowledge are shared among individuals, mainly in the form of Learning Objects and are interpreted as modules of knowledge that arise as a result of the analysis and division of knowledge into "objects" [29].

Development of university-class repositories is influenced by many factors. In that sense repositories have become a bridge between the rapid outdating of domain knowledge and the time of reaction to these changes, giving the possibility to quickly adapt and update the didactic materials. A well-developed repository assures good quality of the learning process. In the evaluation of the universities rankings (eg, Webometrics Ranking of World Universities) repositories are becoming the source of new perception regarding the quality of the entire university.

6.2 Present-Day Status of Knowledge Repositories Adapted to Learning Process

A new paradigm applied in e-learning systems described by the SCORM 2004 standard [40], assumes creating a knowledge repository containing certain domain knowledge divided into distinct knowledge objects (Learning Objects – LOs). The didactic material intended for certain student is developed with creating a sequence of LOs. That fact poses a few significant research problems:

– What should the knowledge repository structure be like in order to facilitate a high-quality description of given domain and at the same time to allow

P. Różewski et al.: Intelligent Open Learning Systems, ISRL 22, pp. 121–150.
springerlink.com
© Springer-Verlag Berlin Heidelberg 2011

adapting the knowledge relayed to the student depending on the
education (e.g. the knowledge depth level) and his/her cognitive features (e.g.
cognitive learning style [26])

–   Development of a method of constructing personalized LO which, as an
    autonomous knowledge model, has to satisfy certain education objectives and
    simultaneously ensure high-quality knowledge, i.e., a context secured in the
    knowledge already owned by the student
–   Development of a knowledge repository management system that describes
    the roles of actors (such as an expert or a teacher), defines the frames of their
    cooperation and the tools that the actors can use to manage the knowledge
    repository, including their scope of operation

The indicated issues have been already analyzed in literature in the context of e-
learning systems. In [25] the knowledge model is a structure of rules described in
New Object-oriented Rule Model (NORM).The authors propose a method
establishing the content of LO in the form of a knowledge class based on the
object-class approach. However, creating a knowledge repository on the rule-
model makes identification of domain and controlling their capacity more
difficult. In [46], the knowledge model has a form of a table called Knowledge
Construction Model. The authors have proposed a method of defining the order of
LO in the form of a knowledge element. However, the structure of LO was not
specified, nor were the criterions defining the content of a LO. In [44] an approach
to building LO has been proposed, in a form of Intelligent Reusable Learning
Components Object Oriented (IRLCOO). The defined knowledge model utilizes
an approach based on a concept graph. In order to define the sequence of LOs
delivered to the student, a multi-agent approach was applied. The method does not
specify the way of defining the size of LO. Moreover, the method of modeling
knowledge with the conceptual graphs cannot model procedural nor fundamental
knowledge successfully. Unfortunately, none of the propositions fulfill entirely the
requirements of the SCORM standard, such as reusability, accessibility, durability
and interoperability of didactical materials and environments of e-learning system.

The issue of LO development has been intensely researched by IT companies
and predominant computer vendors like IBM, Cisco Systems, Adobe and many
others. According to [39], the industrial effort is primarily devoted to content
repositories, learning systems, and authoring tools. For the IT sector the Cisco
Systems has made the biggest effort in terms of LO management. In [4], Cisco's
Reusable Learning Object Strategy (RLOS) has been presented. The RLOS
combines Reusable Learning Objects which include: overview, summary,
assessment and five to nine Reusable Information Object (RIO). According to [4]
each RIO is built upon a single education objective and combined content items,
practice items and assessment items. The RLOS can be seen as a complete
management tool for LO, which is successfully implemented in the CISCO
Networking Academy space. However, the RLOS is missing the procedure for
knowledge management and whole knowledge manipulation depends on the
Subject Matter Expert and the Knowledge Engineer. Other companies focus rather
on their product LO's compatibility. The LO economy is supported by the market
leading learning systems like Blackboard (www.blackboard.com), WebCT

(www.webct.com), TopClass (www.websystem.com), Lotus LearningSpace, IBM LMS (www.ibm.com) and Moodle (http://moodle.org/). The reason standing behind it is that the companies follow industrial standards which continually keep missing the knowledge aspect of LOs and concentrate only on technical issue, like LO's metadata or communication.

6.2.1   The Relation between Knowledge Repository and Knowledge Base

Speaking in the context of knowledge repository-base information systems, let us introduce the following definition of the concept of knowledge. Knowledge is interpreted information that is structured enabling performing the process of reasoning about a researched object. In terms of education, knowledge needs to be oriented towards a particular teaching process objective.

Merging the features of asynchronous learning mode with those of open learning presents considerable requirements for structure and contents of didactic materials conforming to an assumed cognitive model [23,47]. The distance learning material in the form of conventional media-rich websites are characterized with precise and detailed structure and restricted context. Monolithic character of the contents on the one hand and complex reference links on the other impose limitations on didactic materials reuse in a different teaching scenario. That leads to increased cost of their production and shortens the time span of effective use of a single course unit.

A database constitutes a deterministic behavior model of objects and their specific processes in form of a data model of a given structure. Data in the database reflect the values of particular process' parameters and objects in an observed time span. In terms of the Open Distance Learning System-class systems the databases can, for example, offer a means for storage and distribution of administrational and personal data.

A warehouse is a unique kind of database. Gradually supplied with data, it reflects different aspects of an analyzed process/object. It is based on deterministic multi-dimensional data models enabling to perform retrospective analysis of a process execution, its runtime, and behavior of the object on a given time span. Utilization of technologies such as OLAP (Online Analytical Processing) offers mechanisms for conducting interpretation of the accumulated date according to the predefined analytical dimensions. That means that in case of data warehouses we consider information that is processed from the accumulated data. Using OLAP we can acquire information from that data in forms of diagrams and charts that can visualize a trend of surge/decline in course popularity, various educational offerings, student's social status and other kind marketing information.

A knowledge base is founded on a proper knowledge model (rule-base, or semantic network, or others) and it relates to strictly defined subject and is designed to deal with strictly defined class of problems. Using the knowledge model it is possible to define the principles of object behavior or process execution in given conditions. Moreover, knowledge bases can absorb and generate new knowledge as a result of interacting with their users. Using expert-psychologist's knowledge, it is possible to define student's motivational model, or to detail the rules of test selection. Knowledge bases may also be used for the needs of skills acquisition, for example of those specific to using specific hardware.

Joint use of the database, warehouse and knowledge base information systems detailing the same object/process matched with a predefined subject constitute an integrated deterministic model that is capable of: defining current values of object, or process state within a specified snapshot (information generated by the data warehouse) and observing mutual dependencies and influence of the parameters on the given object/process (knowledge contained by the knowledge base).

Both data, information and knowledge related to the same researched object/process can serve as a platform for development of formalized model of management systems for that object with a proviso that all of the concepts used in each of the systems are: sound and semantically consistent, unambiguously interpreted by both the engineers and the users of the system, and all of them match a common ontology (i.e., belong to the same ontological model) and share a common mechanism of creating, supporting and editing of the ontology.

Knowledge repository is intended for representation of philosophical, scientific, scientific-technical, science-technological state of a given domain. The elements of the repository, similarly to the data warehouse, are dynamically provided with data, changing their semantic depth with respect to the teaching objectives. The repository is intended for distribution and sharing the elements of domain knowledge to address the primarily the needs of the educational objectives. In the open systems of distance learning, the repository is a place for storing the Learning Objects that are interpreted as knowledge units as products of analysis and partitioning of the given domain into knowledge segments. The characteristics of the Learning Object formed using the standards of distance learning (for example such as SCORM [40]) provides opportunity for Learning Object reuse in varying distance learning environments without the necessity of conducting its redesign and re-coding.

The range and model of knowledge representation in the knowledge base are predominantly aimed at addressing a certain problem. However, the knowledge stored in the repository should primarily serve the purpose of shaping of the cognitive process and development of creative capacities of students. Ignoring that practically eliminates any prospects for attaining high level of competencies by those individuals. When we deal with the overall problem of development of the open distance learning system, we may conclude that each of the aforementioned elements has its role in that class of systems.

Addressing the problem of development of coherent method of domain knowledge modeling in distance learning systems is a basis for effective application of the knowledge repository concept accepted in the LMS/LCMS-class systems. The most significant element of that idea is modeling of extracted knowledge unit in a form of the Learning Object. Its contents and structure is defined with the XML language. The repository is intended for distribution and sharing of the domain knowledge units for educational purposes. The method of domain knowledge modeling in the context of Learning Object structures should meet the following two requirements:

- The domain knowledge model designed for learning needs is a result of cooperation of different categories of experts, namely domain experts, knowledge engineer and teacher/tutor
- The contents, capacity and depth of knowledge contained by the Learning Object should be equipped with a mechanism for adapting it to different learning objective

A more computer-specific look on the Learning Object has been discussed in [23,47], whereas the idea of the knowledge repository based on the utilization of the Learning Object can be found in [23].

Analysis of possibilities to meet all of the presented requirements leads to following conclusions. Firstly, since it is quite difficult to devise entirely automated methods enabling us to develop of effective Learning Objects, it seems necessary to develop an interface allowing for structure and contents visualization of domain knowledge. That interface would primarily address the needs specific of knowledge engineer and expert. Such language should combine elements of verbalization of elements with limited semantics and syntax. Secondly, since all of the discussed environment is analyzed in the context of information systems, it is necessary to develop certain formal framework enabling us to effectively manipulate the structures of knowledge in the digitized environment.

The presented functions should be satisfy the requirements of one of the modules implemented in the software the knowledge repository will be implemented with. Availability of such module should facilitate the organization of work related to development of contents and structure of didactic materials as a first stage in the process of intangible production. However, it rather difficult to pursue and expectation that the degree of automation at that stage will be as high as in the second stage of development of didactic materials considered as a final product of electronic edition of those materials.

6.3 The Concept of Learning Object

Since the beginnings of distance learning systems, a concept of appropriate method of didactic materials creation has been pursued. A traditional approach is primarily concerned with development of entire course bottom-up and is considered to be uneconomical. As it has been explained by [10], the total cost of developing a typical distance learning course virtually in every occasion include costs of:

- service performed by a domain expert (preferably with pedagogical background),
- network specialist,
- costs related to investigating and regulating copyrights issues,
- official acceptance by academic community (following an official evaluation)
- additional administrative costs.

It is likely the costs will increase if we opt for application of rich-media contents and other sophisticated IT technologies. A monolithic distance learning course is built from scratch and is based on conventional academic course or properly designed course book. The course is offered to limited group of students participating in a limited time span. Any distance learning course that fits this description is guaranteed to be more expensive in development in comparison with the conventional counterpart performed in a traditional form.

As an outcome of scientific research performed in the fields of cognitive science and pedagogy, an idea of application of knowledge modules called Learning Objects has been developed. The concept of Learning Object is presented on figure 6.1. The discussion of that concept is best to conduct using a metaphor of blocks. It has been noted that conceptually identical knowledge modules can be discerned among many courses of the same domain. For example, different courses of mathematics contain the topic of cosine function that is the same regardless of the studies, the same both for architecture and computer science. Therefore it seems reasonable to partition the knowledge of study of a given domain into Learning Object modules. The resulting "blocks" can be interconnected leading to creation of variants of the principal (and core) course [23,47].

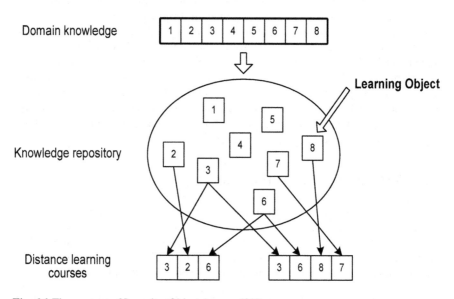

Fig. 6.1 The concept of Learning Object (source [23])

The concept of sharing and reuse of knowledge (not limited to education) using computer technologies and knowledge base emerged in the 1990s [31]. Instead of initiating building and filing the knowledge base on every iteration, a significantly better approach can be achieved through collecting reusable knowledge modules which could be used for effective composition to final form of knowledge to be distributed as per individual requirements. Not unusual for that kind of scenario, some effort would be required for development of initial knowledge base. Later on the costs of deployment are only recorded for gradual expansion of the stored knowledge, which can also prove to be a form of goods subject to commercial transactions.

6.3.1 Learning Object Approach Discussion

The Learning Object economy has become a yardstick for organizing both educational infrastructure organizational unit and its resources [7]. Basing educational activities on the Learning Object repository including a new role of a teacher in the learning-teaching sequence leads to major change of economical factors involved in the learning-teaching process (see figure 6.2). As a result of that, standard educational procedures and blueprints of education are redesigned. The central point of educational institution budget shifts towards initial stages of development of an education offer for students. Any corrective and editing actions upon conducting a course are very limited and usually cost-prohibitive, hence the development of educational materials becomes increasingly important and so is their adequate formation, namely using knowledge about the student entity. In that context, the predominant element of any given organization is the repository of Learning Objects and student profile database. In the long run, the greatest parts of the overall costs are borne with reference to creating and maintaining adequate quality of the Learning Objects repository. The major advantage, however, can be seen in reduction of the costs specific to course development process. That strategy based on use of Learning Objects and their inherent standards enable to conduct an extensive exchange of didactic materials with other organizations. That is considered particularly beneficial as it provides opportunity for deeper exploration of educational possibilities, such as the complete operational research course material prepared by a group of most prominent European experts [22].

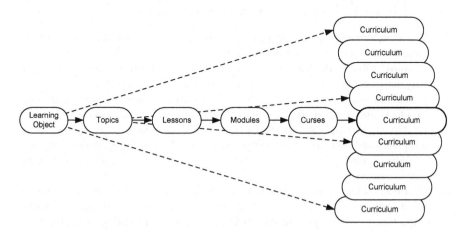

Fig. 6.2 The model of new economy in the context of the Learning Object application

The Learning Objects methodology is considered to be particularly suitable for asynchronous learning. Examination of present-date situation of distance learning in Europe indicates that at the current situation of educational market it is not economically advisable to apply synchronous methods in contexts other than those of highly specialized nature. That situation is mainly caused by average telecommunication infrastructure and rather low level of financial stability of typical recipients of that particular product. Many educational institutions are already prepared for distributing services of synchronous distance learning. The infrastructure that has already been developed is mainly used for supporting educational offerings of a given institution by means of inviting guest speakers to host and conduct lectures.

It is reasonable that since the course textbooks are prepared by few designated and highly successful researchers, therefore a similar strategy should be accepted with reference to the Learning Objects development process. Similarly, such objects prepared by designated experts could be made available in open repositories. Upon developing a course a tutor could resort to Learning Objects that have been previously developed. Almost instantly we encounter claims that each educational scenario is different and the necessity of unification that the debated approach suggests is likely to harm the overall quality of the teaching process. However, as it has been indicated by [41], the tutor interprets specified knowledge portion, performs creative identification and attributes specific context. The textbooks are interpreted by tutor during lesson, whereas a Learning Object is the target form of knowledge in distance learning scenario. The role of the tutor and one's pedagogical practices become changed. Exact repercussions of application of Learning Objects in distance learning, including adequate pedagogical practices, have been investigated by [6,38].

According to the Learning Object concept, courses are not thoroughly developed from beginning to end, but they are more likely to be composed of previously developed Learning Objects. That very change influences many factors. They can be defined according to the research of [15]:

1. The development process is primarily concerned with identifying Learning Objects, not on development of courses or arrangement of lessons: the Learning Object designer looses some part of controlling end material. It may happen that he/she will be placed within the surroundings of Learning Objects of much poorer quality. Countermeasures to such situation need to be in place and they mainly base the development of Learning Objects on predefined formats and templates so that it could improve its quality. The templates and formats arranged by experts enable to refer upon development of didactic material to guidelines and factors tested and deemed effective in real situations.

2. Learning Object is developed with several applications in several contexts in mind: the role of context in learning-teaching process is of paramount importance since its precise specification may lead to limiting its prospective use in other courses. A solution for that problem has been addressed by many individuals in research projects - also identified and debated in other parts of this book. It is worth to mention a proposal presented by [27] dealing with the

problem of specification of base (and suggested) context for Learning Object accompanied with guidelines on adapting of a specified Learning Object to a given context and required by a given course.

3. Contents and their substance is separated from data representation in order to facilitate the process of adaptation to the recipient's format: state-of-the-art didactic materials are created in manner preserving separation of the material contents from mechanism of their representation and visualization. The contents are encoded with computer standards based on extensive languages, for example XML. Forming the Learning Object with accordance to the course specification is limited to manipulation of visualization parameters, for example in CSS files or XSL (standing for XML Stylesheet Language). Not only does that separation permits adaptation of the Learning Object to particular courses, but also to various methods and formats of visualization, for example required by smart phone and netbook devices.

4. Basing development activities of the Learning Object using educational and IT industry-specific standards preserves portability: acting with accordance to standards accepted upon development enables seamless portability of a given Learning Object between different Distance Learning platforms. That in itself leads to independence from hardware platform and operating system. That leads to further used of resources available worldwide (what is the case, for example, when OpenCourseWare platform is used). However, that approach may lead to necessity of determining solutions for problems related to cultural and social backgrounds that occasionally prove to facto in upon transferring knowledge between countries and continents.

5. Learning Object should be defined using meta-tags: speaking in the context of large repositories rich with contents, it is necessary to provide a mechanism for indexing and search. Use of tags and metadata-related standards enables to define each Learning Object instance using parameters such as: format, volume, software requirements, author-specific data, copyright issues, version number, didactic character and others. Core metadata of the Learning Object has been researched extensively by many organizations and that has led to creating several e-learning standards such as IEEE LOM and DublinCore. According to [35], the standards of Distance Learning exploiting metadata languages are expected to address the problem of presenting: the contents of a Learning Object, its purpose and requirements that should be met in order to successfully apply the given Learning Object.

6.3.2 General Pattern for Development of Learning Object

Analyzing general structure of the Learning Object presented on figure 6.3, it can be seen that the scheme makes use of Distance Learning standards and literature of particular subject and constitutes an attempt to aggregate concepts and ideas from standards like SCORM, IMS. Cognitive and pedagogical factors are ingrained within structure of didactical materials, therefore not visible on the figure.

Learning Object

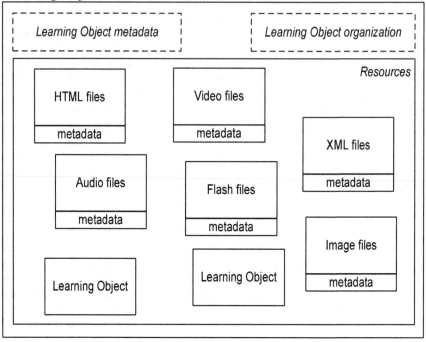

Fig. 6.3 Learning Object structure

The Learning Object is composed of three main components. Each of them has an identified role determined by standards and guidelines of given organization. The components complement one another. Below are the specifics:

1. Resources: they contain files and other Learning Objects that enable to visualize knowledge contained in given object. The resources can be also formed of scripts and active code (for examples please refer to [16]. It is also possible to operate with particular files from copyright policies standpoint. The files contain their own structure, description (based on metadata) and their type determines their volume, means of use, for example video files may put additional load on network, whereas Flash files may require newer versions of Flash players, etc.

2. Learning Object metadata: those elements contain structured and detailed specification of the Learning Object. Based on standards, they usually presents contents of particular Learning Object and lists requirements for the environment, whom it is subject to and who has the rights for its use. It also contains the information the Learning Object is built with. Semi-structural character of the metadata language allows maintaining an autonomy in providing formal description of the Learning Object.

3. Learning Object organization: provides means of using the Learning Object. Certain pedagogical pattern is determined (one or a few) for using a particular Learning Object in different practical contexts. The educational pattern becomes subject to implementation and may serve as guidelines for individuals willing to use a particular Learning Object in a course that they have developed.

6.3.3 The Rules of Learning Object Development

Unfortunately, the standards of conducting distance learning that relate to Learning Object are mainly concerned with issues of portability, interoperability and delivery of universal method for contents description. So far none standards have been identified that would consider development of Learning Object content. Using outcomes of analysis presented in [15] we may review some of discussions treating the rules of developing Learning Object that have emerged in the literature.

First of the rules is based on assumption that Learning Object is an autonomous didactical unit. The author of [30] claims that Learning Object should be autonomous to a degree that would preserve its separation from context. An operation of associating a Learning Object with given context results in adaptation of contents to the specified form (context). Using that approach would lead to attaining more flexibility of didactic material. However, it is considered to be hard to formalize due to use of fairly refined heuristic in development. An entirely different approach has been presented by [10] suggesting pursuing development of Learning Object in a fashion similar to typical training session. Autonomy of Learning Object, presented as in [35], is an effect of development of its contents. Every Learning Object needs to meet a predefined educational objective without a necessity to refer to previous topics which may under particular circumstances never occur in a given course. The proposition made by [27] has raised issues of cultural issues that are strictly tied to the open character of the Learning Object concept. The language of the Learning Object contents should meet the requirement of reaching out to the most of potential recipients. The aforementioned issue of culture-related differences should motivate content creators to maintain high standard of language analysis in every given Learning Object instance. The major cause for misunderstandings due to culture-related differences in Learning Object application is referring to analogies and examples. A variety characteristic for every language forces the content creators to consequent use of terminology, preferably with accordance with standards applicable in given domain.

Second guideline refers to use of educational standards and pedagogical theories upon designing and using Learning Object. Examples those standards are SCORM, IEEE LOM and the group of the AICC standards. In addition, a pedagogical theory requires designers to address two fundamental questions: "What do we want to learn?" and "How do we want to do it". In that sense, application of certain pedagogical strategy provides opportunity for creating development templates and when used they offer a possibility for adapting available didactical material. Learning Objects that will be created in such pattern will be characterized with high level of quality. Unfortunately, a general trend has been indicative of a populated conviction that recent spur of new learning system and new "technological" standards is likely to improve efficiency of the learning-teaching process. As it has been shown by [30] that belief can be easily refuted. Moreover, failing to pay special attention to the development of pedagogical aspect of the process is not likely to increase effectiveness and overall quality of learning-teaching process.

Next concept is related to the problem of preserving a coherent context across a group of Learning Objects. When we connect Learning Objects we also associate specific context with them. It is possible to realize a situation where all the interconnected Learning Objects will be considered as mutually inconsistent, in other words taken of out their context. Therefore, development methods aimed at Learning Objects needs to maintain the context information element so that a student could comprehend the arranged Learning Objects' contents. The author of [27] has proposed several solutions to that issue. Moreover, preserving the additional, exceeding amount of information and knowledge will likely to offer the student an opportunity for creating his/her own conceptualization of a given topic.

6.3.4 Granularity of the Learning Object

One of the most challenging problems related to the development process of Learning Object is determination of granularity of the Learning Object (see figure 6.4.). By granularity we mean a certain volume of Learning Object, or a certain range of knowledge of contained in single Learning Object entity. This problem is important because it is directly related to the concept of providing Learning Object reuse. The objective for that issue is finding a compromise effective enough to offer reuse of a given Learning Object in multiple contexts and yet retain its relative high educational value. On the one hand, preserving the Learning Object's compactness is likely to lead to an increase in its degree of universality thanks to which it will be possible to reduce course-related costs because it will be possible to create a course using previously developed objects. Maintaining Learning Objects of fairly limited volume, what has been noted in [35], allows to increase a capacity of Learning Objects reuse. In addition it also provides, as in [27], to extend a degree of personalization indicating that each of studying individuals should receive individually fine-tuned material. On the hand, however, the arranged portions of knowledge may not meet the identified educational goal due to its rather limited educational value. Granularity of too high degree will lead to a situation where the individual responsible for preparation of the course will be forced to collect and compose a course of excessively great number of Learning Objects.

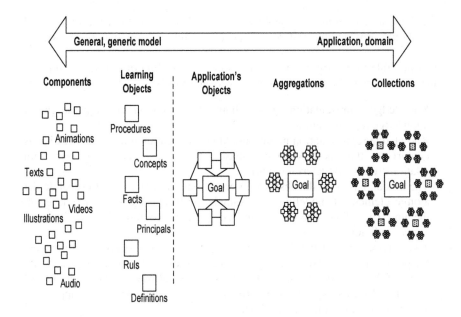

Fig. 6.4 Representation of Learning Object granularity problem

An approach towards determination of appropriate Learning Object granularity oriented towards media is based on the operation of aggregation. The offered mechanism is based on traversing a determined hierarchy and aggregating the hierarchy elements into a unifying Learning Object. As indicated by [18], at the very bottom of the hierarchy are files accessible during the development process of educational materials. The most top-level elements is the course itself, or a learning unit, depending on specific of the course program. The intermediate hierarchy levels are specific to features and specificity of applied approach towards distance learning.

6.3.5 Discussion of Knowledge Representation Models in the Context of Requirements of Learning Object Modeling Process

The base knowledge model arranged with the specifics of distance learning work environment should be based on cognitive structures, i.e. concepts. An atomic unit of the knowledge model should be the concept. The knowledge model integrates in its structure:

- a given conceptual unit,
- operation perform on those units,
- a reasoning system with indispensable reference system,
- identified domain range and that altogether enables an engineer to develop a domain ontology.

Using these assumptions as groundwork for further elaboration, the ontological model should provide coherent communication thanks to exchange of knowledge units.

It is possible to determine the requirements of modeling of conceptual domain model with reference to the following knowledge models:

- Knowledge representation model should be based on visual paradigm of knowledge representation.
- Knowledge representation model should allow separation of particular objects comprising the given knowledge-based system and associate them with relations (ontology determination).
- The model, based on the abstraction operations, should enable to separate the atomic element of modeling, for example a concept. On the semantic level, the unit should be unanimous and adequate to cognitive specificity of a human individual.
- The requirements of the distance learning environment specify requirements for preserving of certain degree of formality in encoding the contents, so that it could be possible to store the results of knowledge modeling in the LMS/LCMS class system and formalize them using one of the applicable standards, for example the SCORM.
- The knowledge model should allow to conduct modeling of fundamental knowledge that is based on concepts and should be stored by the student's long-term memory.
- A formalized system of domain knowledge representation should have the capacity to associate domain knowledge with its learning-teaching methodology and related knowledge elements manipulation language.
- Since one of the main features of ontology is the ability to embed them into a specific domain, the knowledge model should provide means to model a context associated with an element of that domain knowledge.

Identification of knowledge model that is most suitable for the requirements of the conceptual knowledge model for distance learning should be preceded by a discussion on the issued of knowledge modeling. Speaking more specifically, the problem of knowledge modeling is strictly related to the problem of knowledge representation. The latter became extensively researched almost concurrently to the initiation of research activities related to logic. Over the time, the research have gained momentum and reached new added value especially when the concept of artificial intelligence became widely explored by scientific professionals.

As the first step, it is recommended to investigate fundamental methods of knowledge representation presented in the Table 10.1. The rule-based and network-based models constitute cornerstones for further classification of knowledge representation models. Because of their immanent features, those models are suited for different real-world applications.

The rule-based model thanks to its ease of conducting reasoning process is usually a basic mechanism used in the expert system design and engineering. Using facts specified as a set of rules and the modus ponens principle of inference, it is possible to infer either backwards or forward. However, a limitation that is effective for that method is the required strict formalization of the modeled real world in a form of predicates and rules. That formalization may prove somewhat unnatural for an individual. However, due to its rather mathematical specifics it is exceptionally well suited for implementation in a computer system.

The network-based model is premised on an entirely different notion. Knowledge is represented in a form of some identified structure, which architecture determines meaning of particular elements of the network-based knowledge model. Knowledge system of network-based knowledge models is based on concepts in accordance with the human cognition apparatus. That approach can be characterized by greater degree of modeling flexibility that is why it is essential to maintain utmost precision upon conducting domain knowledge identification. Lack of precise modeling formalisms may lead to emergence of difficulties with computerized analysis of knowledge systems in development that are based on semantic networks. One of the possible solutions to rectify that issue is leveraging the solution proposed over 100 years ago by Charles S. Peirce that is based on a network model formed using existential graphs [17,36]. Existential graphs are graphic and symbolic knowledge representation environment. The proposed symbolic notation is similar to the first-order logic. An example of knowledge representation model based on that proposal approach is conceptual graph [43].

Table 6.1 Main methods of knowledge representation

	Atomic unit	Knowledge formalism	Inference method	Inference algorithm	Applications
Rule-based model	Facts (predicate)	Rules, theorems	Modus ponens	Backwards and forward inference	Knowledge bases in expert systems
Network-based model	Concepts	Relationship between concepts	Inference is based on traversal of the knowledge graph	Inspecting the knowledge network various conclusions are formed	Semantic networks

In the authors opinion the idea of creating Learning Object using domain concepts integration approach requires use of semantic network-based knowledge model. The analysis should primarily deal with the main networked-based knowledge modeling methods: semantic and neural networks models what has been presented on Table 10.2. Differentiation of both discussed methods has been acknowledged on the very beginning of the analysis. Both models have evolved from outcomes of research dealing with the mechanisms of human memory. Semantic networks's primary role has been dealing with cognitive aspects occurring in human mind, whereas the neutral networks are composed using reference to real physical, chemical and biological process of that very mind and therefore can be considered as an attempt to mimic and reflect its structures. The approach used in the case of semantic networks is originates from the works of cognitive science. However, the approach comprised of modeling several processes using neural networks is often referred as connectionism. [2]. Using the aforementioned subject of discussion, including the concept of ontology, yields to limiting the scope of further analysis to the selection of models that are based on semantic networks.

Model of semantic networks serving as a reference model of network knowledge representation is a graphical representation of knowledge that is based on cognitive approach towards reality. Semantic networks were created as an outcome of the research conducted by Quillian [34] that dealt with structure and architecture of human memory. In the most rudimentary form of semantic network its nodes reflect objects, whereas connections between nodes represent semantic relationship (dependencies between objects). Semantic network is always in place where information is presented by means of a graph, what has been indicated by [14]. For many years psychological research within the area of cognitive science have become stimulants for development of semantic networks. As shown by [13], knowledge models that are based of semantic networks are both well comprehended by human individuals and it is also possible to adapt semantic networks to automated processing. Some semantic networks have been specialized to model cognitive mechanisms of human being, such as mind map. The main purpose of the development process of different research group was reaching efficiency of computerized knowledge representation and processing, for example the knowledge grid by [48].

Table 6.2 List of network-based knowledge representation models

	Determinant of modeling process	Means of knowledge representation	Atomic unit	Structure	Mechanism	Basic objectives
Semantic network model	Metaphor of semantic memory	Based on relations between nodes	Concept (node)	Graph	Abstract representation serving as instrumentalismic model	Visualization of semantic dependencies
Neural Network Model	Metaphor of human brain activities	Based on numerical thresholds (biases) associated with inner connections between processing nodes	Neuron	Layered graph	Dispersed representation of concurrently processing transformation (calculation) elements	Classification problem

The concept of semantic networks is apparently too universal and as a result overtly sophisticated. Its fundamental concept have evolved and adapted to specific applications. Understanding of the terms of concept and relations have evolved which are the basic components of every knowledge model. The concept is a logical unit of knowledge model, whereas the relation is a mechanism to create more sophisticated structures. Role and functioning of both the concept and the relation have been redefined with accordance of given application.

The concept specified using so called Formal Concept Analysis is considered as a unit of mind processes, mainly the process of thinking. Forma Concept Analysis is a mathematical knowledge model that is founded on philosophical comprehension of the term of concept researched by [8,12]. The formalization is founded on the conception that human intellectual and communication processes are always embedded in specified context that determines meaning of particular concepts. The analysis primarily deals with concept formalization through achieving formalization of the context.

In the model of conceptual maps proposed by Novak [32], a concept is defined as a perceived regularity among events or objects nominated by a specific label. The dependencies between concepts are created by forming assertions associating several concepts (two at minimum) into a form of comprehensible meaning. Such assertions are called semantic units. The concepts are formed into hierarchy in which top level is occupied by most semantically-generic concepts, whereas the bottom levels are occupied by most precise detailed and adequate concepts of a given domain. The contents and structure of the map is determined by so called cognitive economy. That form of economy can be considered as a semantic organization of location of every concept in the conceptual structure. In essence, it is primarily concerned with a concept - an aggregate of other concepts - is located in the semantic hierarchy higher in order to span other concepts that relate to it. Other important factor to be considered in the conceptual maps is the ability to partition and group knowledge elements called knowledge clustering. That rule stems from the frame theory and basically is suited for providing formalized description of abstract or real object using certain portion of a conceptual map. The idea of conceptual maps is based on the psychological theory proposed by David Ausubel [3] which assumes that learning is a process of connecting new concepts to those cognitive structures that have been already organized in a student's mind and as such has been defined as meaningful learning. Hence, it is the main presumption upon which the conceptual maps have been established.

Not all of the connections between concepts and relations may seem reasonable, especially at a first glance, what is difficult to be formally assessed. In case of the conceptual graphs the correctness of the connections is ensured by canonical form as a class of the graphs that have been created using those canonical graphs. The problem of context identification in the contextual graphs has been determined by use of the canonical form structure presented by [9,11]. The canonical form is a starting point for creating a conceptual graph specified in a given context. The latter element imposes semantic limitations and provides necessary information related to particular domain of application. The canonical form is defined by the following elements: the hierarchy of concepts (ordered with the relations), hierarchy of identified relations, set of individual tags and mapping function that bind the elements of the set of individual tags with the hierarchy of concepts and relations. All conceptual graphs that are successfully generated on the basis of the canonical forms are called canonical graphs and are based on the canonical form's elements and are valid (sound) for the given domain. A model of the conceptual graph proposed by John F. Sowa [42] includes certain number of operations [11] that yield to obtaining new canonical graphs using the following canonical operations: copy, restrict, simplify, join.

Many of knowledge representation methods have developed unique means of determining context. In the case of semantic networks specialization of semantics is not available and that fact has been proven by [28]. Semantic networks are usually effective in presenting meaning of objects without imposing formal restrictions on the reasoning process. It leads to mounting of difficulties related to conducting computer analysis due to impossibility of deterministic declaration of analyzed knowledge space, i.e. the context. It is the context that identifies areas of knowledge that proves useful in interpretation, determines limits effective for reasoning process and all other knowledge-specific operations for knowledge defined by the knowledge model.

In the formal concept analysis, knowledge is encoded in tabular form. As such, it represents polysemantic context [45]. In order to obtain an unequivocally polarized concepts, it is essential to perform transformation onto a single-dimension context (table) using the operation of conceptual scaling. In that context, is required to properly identify which attribute is crucial and how it should be transformed onto a single-dimension area. Formally, context is defined as: $K := (G, M, I)$, where G - set of objects, M - set of attributes, and I is a binary relation between objects and attributes: $I \subseteq G \times M$. In many cases, it is mandatory to analyze multi-context instead the single-dimensional context. In that case, it is necessary to synthesize of several identified contexts that have the same common denominator into the form of singe-dimension context.

Context defined by topic maps is considered as allocation of knowledge within a structure of an organization's resources. Knowledge is not an abstract value and is instantiated by files, or documents. Topic maps [19] are composed of topics and relations between the topics and resources. Every topic associated with relation has unique role that is defined by a role type. A typical topic map diagram is divided into two parts. The first part can be called as topic's domain, whereas the other as the resource's domain. Both topics and associations between topics make up a semantic network that spans the resources and related with those using the relation of existence. In that sense, context is defined on the levels semantics and the resources it accommodates.

Analysis of possibilities for expanding context in a knowledge model to accommodate both abilities and skills has been performed upon research in the field of skill maps [1]. They are based on top of topic maps and have an additional layer of skills. The origin of that knowledge representation method is insufficient capacities of topic maps for analyzing skills and knowledge of an employee conducting search in a knowledge repository represented by topic maps. Third layer stores personalized information on abilities and skills of an employee and also the information about how they have been acquired.

Knowledge visualization / representation model requires maintaining high degree of communication and interaction between the model and professionals using them who frequently do not know enough about the issues of knowledge modeling. To provide an expert with satisfactory work conditions leading to effective outcomes, that environment should have straightforward rules concerning the knowledge model and effective knowledge elements manipulation language. That environment should properly interact with psychological factors of learning process.

A model that appears to be adapted sufficiently enough to accommodate the issue of visualization of the Learning Object modeling is conceptual maps. Semantic relations that interconnect particular concepts are directly derived from an approach towards teaching particular knowledge and are a representation of means of inference in a given expert's domain. Conceptual maps allow for thorough determination of the domain which the modeling is performed with. That leads to attaining certain level of flexibility of manipulating concepts. The visualization of a particular domain portion is directly transformed into and educational approach towards teaching that knowledge eliminating a need for indirect transformations.

6.4 Functioning of Knowledge Repository

Distance learning system equipped with knowledge repository absorbs management capabilities that include both the Learning Objects and those elements on lever levels, such as knowledge portions (chunks). That ability of working with knowledge contents cross many levels offers greater flexibility of the system thanks to attaining greater precision in the knowledge modeling process. Location of the knowledge in the knowledge repository from more than one domain (using only a single paradigm) enables us to create a multi-contextual material that integrates in its structure various amount and contents of knowledge that is unified by the author of that material.

Use of knowledge engineering methods for distance learning methods affects distance learning methodology concerned with creation of didactic materials. The actors participating in that process change as well as their roles alter. All of those changes need detailed analysis in the context of pedagogical factors and requirements of the learning-teaching process. Some relations and procedures cannot be neglected, for example upon maximizing the automation degree of that system. Similarly, a new perspective on the aspects of the process, for example specific for cognitive science, allows to reorganize of given didactic scenario. As a result of that analysis two stages have been introduced as part of the proposed method [23,47] that need to be provided with a solution. The stages are:

- Concepts network creation algorithm (the main activities are identification and representation of a problem domain)
- Compilation of didactic materials algorithm

The first step of the concepts network creation algorithm is identification of the concepts pertaining to the domain. That should provide an opportunity of creating a concept dictionary that includes all of the identified domain concepts. As the next step, the properly identified concepts are provided with a structure. Using the domain modeling methods, a semantic network of relations is created that reflects the conceptual domain model. The process of extracting knowledge from a domain expert is equal to the process of knowledge modeling (e.g. defining the semantic structure). The final outcome of that stage is a problem knowledge model (i.e. concepts network) with accordance to computer structures and located into the repository conforming with requirements of ontology.

The second stage is concerned with creating a dedicated Learning Objects sequence that is relayed to the student. Using the representation of a problem defined in the teaching process objective, the selected methodology of teaching and the student's profile a reorganization of knowledge structure is performed. Since the most desired situation is the one where the student absorbs knowledge in systematic manner, the computer system should be provided with sequentially-organized didactic material.

The proposed system functions with accordance to the scheme presented on figure 6.5. The solution is embedded in structure of the LMS/LCMS system. Both stages of the proposed solution require participation of the following actors: the student, the teachers (tutor) and the knowledge engineer. Figure 6.5 presents the problems specific to that system referring to two identified layers: ontology layer and the SCROM layer.

One of main project determinants of the knowledge repository system development is maintaining knowledge sharing and reuse. That task is performed on the level of ontology. The knowledge engineer develops a domain model formalizing conceptual domain model into ontology. Upon creating the latter component the engineer utilizes the portion of knowledge absorbed from the domain expert. Another viable knowledge source can be external knowledge bases, such as those in form of global repository of Learning Objects, or the tutor. Due to flexibility involved in the language of knowledge elements manipulation, the expert can actually perform the task of ontology development on his own. Knowledge is stored in the distance learning system in a form of ontological structure of conceptual domain model. However, such organized form of model may prove not to be useful when we relate it to the recent applications of distance learning based on the SCORM standard. Therefore, the knowledge repository system is implemented with the method enabling transformation of the ontological form to a modular equivalent represented by Learning Objects. Semi-automatic mechanisms, algorithms and transformation procedures perform the task of didactic materials compilation using information stored in the student's profile and participation of the tutor who is actively engaged in development of incoming Learning Objects preserving the teaching objective of a given course and using the guidelines of specified teaching methodology.

An additional comment is required for knowledge base structure that has two dimensions. Taking into consideration the task of knowledge modeling under defined ontology - the knowledge base contains structure defining domain knowledge that is based on set of concepts and their relations. The ontological component offers versatile manipulation capabilities of manipulating concepts and relations. In the distance learning system ability of communication using standard language and structure offered by the SCORM standard are of significant importance. In that context, the SCORM standard's interface becomes the second layer of the debated knowledge base system. It changes the ontology structure into a set of interconnected Learning Objects using an appropriate mechanism of transformation. The Learning Objects can be manipulated using typical editors and tools specialized in editing and developing distance learning didactic materials.

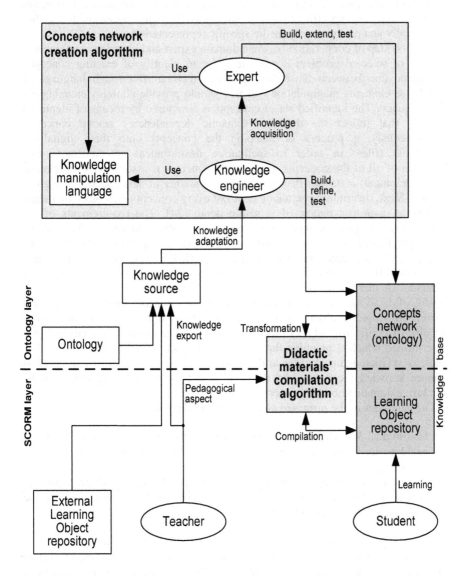

Fig. 6.5 A diagram of dependencies occurring between actors of the knowledge repository system (source [23])

6.4.1 Task of Specification of a Problem Domain

Analysis of tasks that need to performed upon specifying a problem domain are based on the publications of [47]. The initial stage of preparation of didactic material based on the concept of Learning Objects is designed to identify and formalize knowledge specific to given domain. Figure 6.6. presents an algorithm determining sequence of steps for creating conceptual domain model. It contains concepts that are specific to given domain. They are associated together semantically and provided with media-specific representation.

The first step of cooperation between domain expert and knowledge engineer is analysis of selected problem domain in order to identify of existing concepts. Communication between those two agents is maintained through language of knowledge elements manipulation and that should provide contents unambiguity and adequacy. The identified set of concepts is structured by means of identified relations that reflect the domain's semantic dependencies among concepts. Simultaneously, a process of mapping the concepts onto their digitalized equivalents (files) in order to facilitation development of multimedia-rich definition of all of the concepts. Each of the concepts is associated with specific visual (graphical) metaphor that in itself is a container of data specific to a given concept. Next, determining semantic depth of every concept takes place by means of creating a proper matrix of semantic depth [24]. The requirements of the distance learning process impose restrictions in visually representing the concepts into digitalized form. The repository stores data set specific to each of the concepts including metadata (such as date of creation, author, etc.) together with a collection of files relevant to the concept metaphor. The expert has the prerogative to decide whether the metaphor is adequate to a given concept. The expert is also capable of deciding whether the semantic depths of a concept should be broadened keeping in mind problem domain-specific knowledge. Once a satisfactory outcome of verification of all of the concepts has been achieved, the next process is performed which is mapping the concepts onto their visual representation. The whole is verified again in order to determine if the identified set of concepts has spanned knowledge contained by the domain and identified by the expert as the conclusive domain knowledge specification.

The outcomes of the proposed algorithm are the following:

-   Ontological knowledge model (representing the conceptual domain model) enabling us to map the concepts onto their visual metaphors stored in the repository.
-   Thesaurus containing definition of the concepts. It is a designated structure created on the basis of repository resources and is a projection of the conceptual domain model.

Figure 6.7 presents an example of conceptual modeling which objective is development of a conceptual domain model. A mind of the expert (a) is represented by a virtual set of concepts (b). According to the presented conclusions, a network of concepts is the key element of that model. Associating

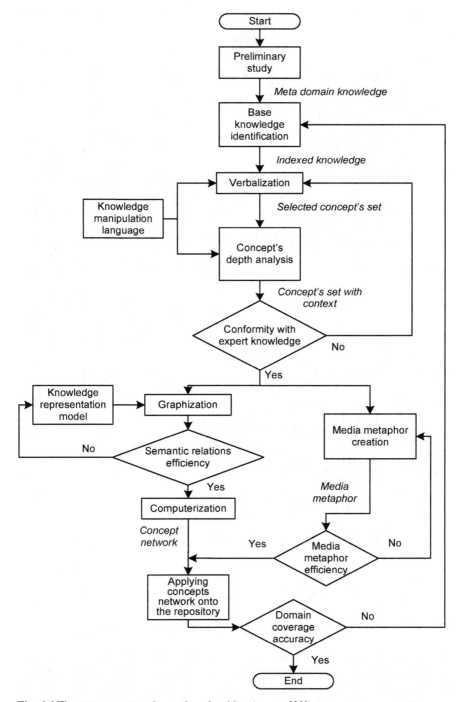

Fig. 6.6 The concepts network creation algorithm (source [23])

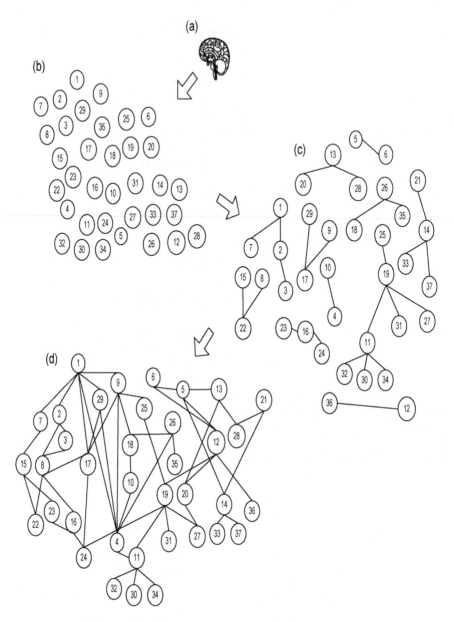

Fig. 6.7 Specification of domain knowledge based on conceptual modeling

at first the concepts into unsophisticated forms (c), a finer identification becomes possible for that network. The result of the process is a conceptual domain model (d) developed by a domain expert with assistance of a knowledge engineer using a determined method of knowledge extraction and its further organization into structures.

6.4.2 Task of Didactic Materials Compilation

The algorithm of didactic materials compilation (please refer to figure 6.8) performs adaptation of the didactic materials to the expectations of an individual student with accordance to the standards in place as well as educational and cognitive factors of the process. The outcome of the algorithm is a sequence of Learning Objects that is relayed to the student through distribution channels of the LMS/LCMS system and Internet. The algorithm utilizes the following elements: conceptual network, repository, student's profile and didactic objective. Using the conceptual domain model and also the information about student's actual advancement stage - base knowledge – and limitations of the didactic process (didactic objective), the knowledge network becomes organized in a layout that best corresponds to that base knowledge and their associations with different other concepts. In that context, it is mandatory to eliminate certain concepts due to the didactic objective that identifies the amount of knowledge the student should absorb in a given domain. The outcome of the organization process is a hierarchical network that contains on its top level the base knowledge concepts – the concepts that have been already known to the student. The concepts located below that level are considered to be equal didactic goals (knowledge that is unknown to the student). In the next stage the multi-level hierarchical network is decomposed into a set of subgraphs that are transformed into trees. That process may require doubling: a concept that is connected with a relation to a different concept becomes doubled in a manner that it could be located into every possible contexts with respect to the structure of the tree. Each of the formed trees is analyzed by means of the algorithm of concept spanning so that a knowledge portion that could meet the cognitive requirements could be determined. The portion becomes a subject of instruction and as a Learning Object is distributed to the student as an element of learning path in the given sequence. Transformation of the knowledge portion to the Learning Object requires mapping the set of concepts onto the contents of the repository. The algorithm in more details is presented in [47].

Data enriched by a necessary description are analyzed with regards to the computer aspect. Volume of data composing a single Learning Object is investigated. Taking into consideration the physical network's limited bandwidth, accessibility to resources is analyzed as well as the copyright and ownership factors. It is also possible to adapt the didactic material that is being created to individual, cognitive specifics of every student. Next, particular concepts pertaining to given Learning Object are consolidated and traversal of networks of consolidated Learning Objects is determined. All of the elements created in that process are compiled into a form compliant with the SCORM standard. The final product may be analyzed by the teacher (tutor) that can investigate efficiency of the outcome didactical material with regards to his/hers didactic experience in the given subject of instruction. Finally, the course is distributed to the student through the mechanisms of the LMS/LCMS system.

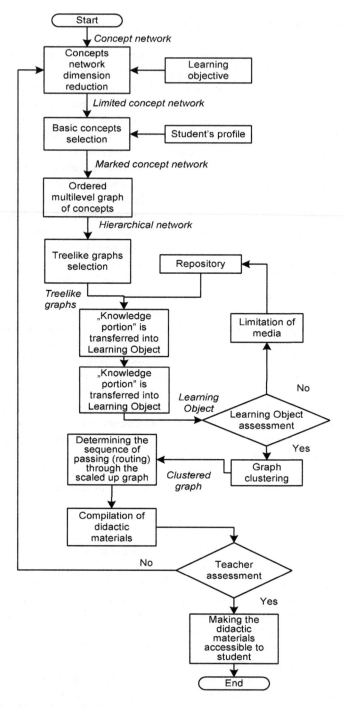

Fig. 6.8 Algorithm of compilation of didactic materials (source [23])

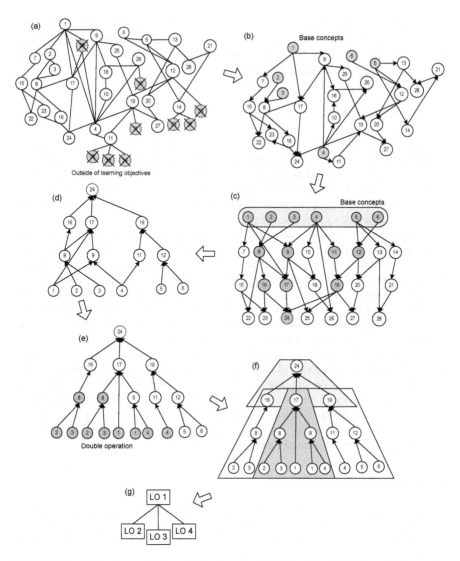

Fig. 6.9 Algorithm of compilation of didactic materials determined on the level of concepts manipulation

The stage of didactic materials compilation that is performed on the level of concepts manipulation is presented on figure 6.9. Taking into account the didactic objective that corresponds to the knowledge a student should absorb, it is possible to reduce the conceptual model (a) as in the example where the concepts 29, 30 and 31 have been reduced. Using student's profile, identification of concepts that pertain to the student's base knowledge (b) are identified, for example the concepts 1, 2, 3, ... In the following step, a hierarchical network (c) is created that is determined on the current mode of teaching (deduction-, or or induction-based) and the relevant subgraphs are extracted that define the student's concepts (d). The formulated graphs are transformed into trees (d, e). A tree is spanned to a degree that corresponds to a single Learning Objects entity (f). The extracted areas are integrated into a form of unified graph (g).

6.5 Conclusions

Development of distance learning systems is related not only to development of more intelligent didactic systems using the student - computer interaction, but also to development of an adequate computer systems infrastructure. Distance learning computer systems have already become a separate software market. It is possible due to existence of appropriate standards that are effective for courses, repositories, didactic systems, tests, and others.

Repositories of didactic materials have become one of the most significant directions in development of computerized distance learning systems. The capital of knowledge required for development to continue has been successfully accumulated. Its multiplication necessitates creating an effective infrastructure for the repository to function properly. The authors of this material have presented definition and classification of distance learning repositories. In addition, we have presented the standards of distance learning that currently support their development and are intended for providing support upon creating global contents infrastructure.

Creating an intelligent method of modeling of Learning Objects requires authors to maintain a high level of personalization. Because each of the students has his/her own cognitive characteristic and his/her own learning style, the basic structure of knowledge repository should offer reduced granulation than the structure of the Learning Object. Conceptual network built leveraging the expert's knowledge and represented through ontological model appears to offer a satisfactory solution for limited form of knowledge storage.

References

1. Abramowicz, W., Kowalkiewicz, M., Zawadzki, P.: Towards user centric e-learning systems. In: Bussler, C.J., McIlraith, S.A., Orlowska, M.E., Pernici, B., Yang, J. (eds.) CAiSE 2002 and WES 2002. LNCS, vol. 2512, pp. 109–120. Springer, Heidelberg (2002)
2. Anderson, J.R.: Cognitive Psychology and Its Implications, 5th edn. Worth Publishing, New York (2000)
3. Ausubel, D.P.: The Psychology of Meaningful Verbal Learning. Grune and Stratton, New York (1963)
4. Barritt, C., Lewis, D.: Reusable Learning Object Strategy: Definition, Creation Process, and Guidelines for Building. Cisco Systems, Inc. (2000)
5. Björk, B.-C.: Open access to scientific publications - an analysis of the barriers to change? Information Research 9(2) (2004)
6. Boyle, T., Cook, J.: Towards a pedagogically sound basis for learning object portability and re-use. In: Proceedings of 18th Annual Conference of the Australian Society for Computers in Learning in Tertiary Education, Biomedical Multimedia Unit, pp. 101–109. The University of Melbourne (2001)

7. COHERE (Collaboration for Online Higher Education Research) Group: The Learning Object Economy: Implications For Developing Faculty Expertise. Canadian Journal of Learning and Technology, 28(3) (2002)
8. Cole, R.J., Tilley, T.A.: Conceptual Analysis of Software Structure. In: Proceedings of Fifteenth International Conference on Software Engineering and Knowledge Engineering SEKE 2003, Knowledge Systems Institute, pp. 726–733 (2003)
9. Cyre, W.R.: Capture, Integration, and Analysis of Digital System Requirements with Conceptual Graphs. IEEE Transactions on Knowledge and Data Engineering 9(1), 8–23 (1997)
10. Downes, S.: Learning objects: Resource for Distance Education Worldwide. International Review of Research in Open and Distance Learning 2(1) (2001)
11. Ellis, G.: Compiling Conceptual Graphs. IEEE Transactions on Knowledge and Data Engineering 7(1), 68–81 (1995)
12. Ganter, B., Wille, R.: Formal Concept Analysis: Mathematical Foundations. Springer, Berlin (1999)
13. Gordon, J.L.: Creating knowledge maps by exploiting dependent relationships. Knowledge Based Systems 13(2-3), 71–79 (2000)
14. Griffith, R.L.: Three principles of representation for semantic networks. ACM Transactions on Database Systems 7(3), 417–442 (1982)
15. Hamel, C.J., Ryan-Jones, D.: We're Not Designing Courses Anymore. In: Proceedings of Conference WebNet 2001, USA, Orlando, October 22-26 (2001)
16. Hanisch, F., Straer, W.: Adaptability and interoperability in the field of highly interactive web-based courseware. Computers & Graphics 27(4), 647–655 (2003)
17. Higgins, D., Heuveln, B., Hatfield, E., Kilpatrick, D., Wong, L.: A Java implementation for Peirce's existential graphs. The ACM Journal of Computing in Small Colleges 16(3), 101–107 (2001)
18. Hodgins, H.W.: The future of learning objects, In: Wiley D (Ed.), The Instructional Use of Learning Objects: Online Version (2000),
http://reusability.org/read
19. ISO/IEC: Information technology – SGML Applications – Topic Maps, ISO/IEC 13250, International Organization for Standardization, Geneva (2000)
20. Xia, J.: A comparison of Subject and Institutional Repositories in Self Archiving Practices. The Journal of Academic Librarianship 34(6), 489–495 (2008)
21. Jones, R.: Giving birth to next generation repositories. International Journal of Information Management 27, 154–158 (2007)
22. Kassanke, S., Steinacker, A.: Learning Objects Metadata and Tools in the Area of Operations Research. In: Proceedings of ED-MEDIA 2001 World Conference on Educational Multimedia, Hypermedia and Telecommunications, Tampere, vol. 1, pp. 891–895 (2001)
23. Kusztina, E., Zaikin, O., Różewski, P.: On the knowledge repository design and management In E-Learning. In: Lu, J., Da Ruan, Zhang, G. (eds.) E-Service Intelligence: Methodologies, Technologies and applications. SCI, vol. 37, pp. 497–517. Springer, Heidelberg (2007)
24. Kushtina, E., Różewski, P., Zaikin, O.: Extended ontological model for distance learning purpose. In: Reimer, U., Karagiannis, D. (eds.) PAKM 2006. LNCS (LNAI), vol. 4333, pp. 155–165. Springer, Heidelberg (2006)
25. Lin, Y.T., Tseng, S.S., Tsai, C.-F.: Design and implementation of new object-oriented rule base management system. Expert Systems with Applications 25(3), 369–385 (2003)
26. Liu, Y., GinTher, D.: Cognitive Styles and Distance Education. Online Journal of Distance Learning Administration 2(3) (1999)

27. Longmire, W.: A Primer on Learning Objects. Learning Circuits - ASTD's Online Magazine All About E-Learning (2000),
    http://www.learningcircuits.org/mar2000/primer.html
28. McDermott, D.: Artificial Intelligence Meets Natural Stupidity. In: Haugeland, J. (ed.) Mind Design: Philosophy, Psychology, Artificial Intelligence. MIT Press, Cambridge (1981)
29. McGreal, R.: A Typology of Learning Object Repositories. In: Adelsberger, H.H., Kinshuk, P.J.M., Sampson, D. (eds.) Handbook on Information Technologies for Education and Training, 2nd edn., pp. 5–28. Springer, Heidelberg (2008)
30. Merrill, M.D.: Knowledge Objects. CBT Solutions, (March/April 1998)
31. Neches, R., Fikes, R., Finin, T., Gruber, T., Patil, R., Senator, T., Swartout, W.R.: Enabling technology for knowledge sharing. AI Magazine 12(3), 36–56 (1991)
32. Novak, J.D.: Clarify with concept maps: A tool for students and teachers alike. The Science Teacher 58(7), 45–49 (1991)
33. Peters, T.A.: Digital Repositories: Individual, Discipline-based, Institutional, Consortial Or National? The Journal of Academic Librarianship 28(6), 414–417 (2002)
34. Quillian, M.R.: Semantic Memory. In: Minsky, M. (ed.) Semantic Information Processing. MIT Press, Cambridge (1968)
35. Quinn, C., Hobbs, S.: Learning Objects and Instruction Components. Educational Technology & Society 3(2) (2000)
36. Roberts, D.D.: The Existential Graphs of Charles S. Peirce. Mouton and Co. (1973)
37. Różewski, P., Ciszczyk, M.: Model of a collaboration environment for knowledge management in competence-based learning. In: Nguyen, N.T., Kowalczyk, R., Chen, S.-M. (eds.) ICCCI 2009. LNCS, vol. 5796, pp. 333–344. Springer, Heidelberg (2009)
38. Ruyle, K.E.: Guided Discovery Teaching Methods and Reusable Learning Objects. The eLearning Developers' Journal (February 3, 2003)
39. Sampson, D., Karampiperis, P.: Towards Next Generation Activity-Based Learning Systems. International Journal on E-Learning 5(1), 129–150 (2006)
40. SCORM: Sharable Content Object Reference Model, Advanced Distributed Learning Initiative (2004)
41. Sosteric, M., Hesemeier, S.: When is a Learning Object not an Object: A first step towards a theory of learning objects. International Review of Research in Open and Distance Learning 3(2) (2002)
42. Sowa, J.F.: Knowledge Representation: Logical, Philosophical and Computational Foundations. Brooks Cole Publishing Co., Pacific Grove (2000)
43. Sowa, J.F.: Semantics of Conceptual Graphs. In: Proceedings of 17th Annual Meeting of the Association for Computational Linguistics, pp. 39–44. La Jolla, California (1979)
44. Valderrama, R.P., Ocaña, L.B., Sheremetov, L.B.: Development of intelligent reusable learning objects for web-based education systems. Expert Systems with Applications 26(3), 273–283 (2005)
45. Wille, R.: Conceptual Structures of Multicontexts. In: Eklund, P., Mann, G.A., Ellis, G. (eds.) ICCS 1996. LNCS, vol. 1115, pp. 23–39. Springer, Heidelberg (1996)
46. Wu, C.-H.: Building knowledge structures for online instructional/learning systems via knowledge elements interrelations. Expert Systems with Applications 26(3), 311–319 (2004)
47. Zaikin, O., Kusztina, E., Różewski, P.: Model and algorithm of the conceptual scheme formation for knowledge domain in distance learning. European Journal of Operational Research 175(3), 1379–1399 (2006)
48. Zhuge, H.: A Knowledge Grid Model and Platform for Global Knowledge Sharing. Expert Systems with Applications 22(4), 313–320 (2000)

Chapter 7
Methods and Algorithms for Competence Management

7.1 Introduction

Due to the constant development of expert systems, broadening of the application area of decision support algorithms, and together with the appearance of new organizational structures, which were defined in world literature as knowledge-base organization, more and more often we come across the concept of competence in the context of information systems.

Processing competences in the environment of information systems requires preparing a set of algorithms and procedures. The first step is to treat competence as a structure which allows for competence description in connection to a specific task and a personalized dimension. In this chapter to approaches to modeling competence structure were proposed. The first one bases on using ontologies and concentrates on modeling the knowledge included in the competence. The second one uses the apparatus of competence theory, what allows for modeling the strengths of relations that occur between individual parts of the competence.

7.2 Nature of Competence

7.2.1 Competence Definition

Many different ways of understanding competence can be found in literature, the authors will mainly base on the [26]. This is the result of the definition being strongly influenced by the context of the studies within which it was discussed, i.e. different fields of science and humanities (e.g. sociology, philosophy, psychology, pedagogy and education etc.). According to Romainville, the French word 'compétence' was originally used to describe capabilities to perform a task in the context of vocational training [23]. Later on, found its place in general education, where it was mainly related to the "ability" or "potential" to act effectively in a certain situation. Perrenoud claims that competence is not limited to just the knowledge of how to do something, but it reflects the ability to apply this knowledge effectively in different situations [22].

P. Różewski et al.: Intelligent Open Learning Systems, ISRL 22, pp. 151 – 176.
springerlink.com   © Springer-Verlag Berlin Heidelberg 2011

The very broad and sometimes even ambiguous meaning of competence has been shown in the review of competence definitions in [26]. A very brief and precise definition is provided by the International Standard Organization in its ISO 9000:2005 standard "Quality Management Systems". The definition that can be found in this document describes competence as "demonstrated ability to apply knowledge and skills". The European Council proposed to define competence as general capabilities based on knowledge, experience, and values that can be acquired during learning activities [5].

The term "competence" is often used interchangeably with the term "competence model" in the literature. The competence model, defined by Mansfield as a detailed, behavioral description of employee characteristics required to effectively perform a task [19], can be used to describe a set of competences connected to a task, job or role in an organization. The form of this description can be of any kind – from a verbal one, to a formal, mathematical model.

Table 7.1 Main competence application analysis (source [26])

Competence application	Job description	Learning outcomes description
Definition	Competence is shown by an actor playing a role. It is an observable or measurable ability of the actor to perform necessary action(s) in given context(s) to achieve specific outcome(s).	Competence is an outcome of a learning process in the Life-long Learning paradigm. It is defined as any form of knowledge, skill, attitude, ability or learning objective that can be described in the context of learning, education or training.
Motivation	The reason for research on the competence concept is the belief of many managers that analysis of organization's structure and resources is not satisfactory. The managers focus on the output of competences.	Open and Distance Learning presumes that students are mobile across different universities and educational systems within the common learning framework i.e. European Higher Education Area. Moreover, the Life-long Learning concept assumes that the student's knowledge can be supplemented and extended in other educational system over the life time. For both of these concepts the well established and transparent method for student's achievement recording is required.
Main characteristics	Competence in the job context is a dynamic system and provides systemic, dynamic, cognitive and holistic framework for building management theory. Moreover, competence explains the formal and informal way in which human beings interact within the constrains of technology, human begins, organization, culture .	Student's competence can be certificated on every single step of training, learning, etc. Competence achievement is mainly based on the cognitive process. However, the competence, like other human characteristics, can be discussed from several different angles: pedagogical, philosophical and psychological.
Standards	HR-XML [12], ISO 24763 [16],	IMS RDCEO [15], IEEE 1484.20.1 [13]
Projects	TENCompetence [33]	European Qualifications Framework (EQF) [8], ICOPER [14], TENCompetence [33]

The analysis of competences is based mainly on visible results. Just as it is the case with knowledge, the concept of competence cannot be directly examined. Competence is a construct concept which is derived or inferred from existing instance [18]. Proper indicators of construct recognitions rely on many different factors: social, cultural and cognitive [6]. The concept of competence has become an important issue for researcher over the world [28,29]. Table 7.1 discusses the competence issue following two main competence applications: job description and learning outcomes description.

The TENCompetence Project – a large research network founded by European Commission through the Sixth Framework Program with fifteen partners throughout Europe [33] – is a good example of the recent research initiatives that provide more and more elaborated formal models of competence. This project is mostly aimed at developing a technical and organizational infrastructure to support Life-long Learning in Europe and focuses its research mainly on the problem of managing personal competences. One of the results is the development of the TENCompetence Domain Model (TCDM) that covers many important issues related to the notion of competence [33]. The TENCompetence project team proposed the definition of competence as "effective performance within a domain/context at different levels of proficiency".

7.2.2 Evolution of the Concept of Competence

Theoretical knowledge combined with proper procedural and project knowledge creates the basis for competence. The necessity of combining these types of knowledge arises during performance of practical tasks, laboratories and projects. For example, the methodology of conducting classes in the frames of a virtual laboratory, apart from other pedagogical rules, should include defining ways of differentiating theoretical, procedural and project knowledge regarding a given topic, domain.

In the context of the presented discussion it is not important whether the theoretical knowledge has an a priori or an a posteriori form. In the discussion the used definition of competence assumes that competence combines three types of knowledge: know what, know how, know why. In this context the competence can be defined in the following way: competence is the ability to use procedural knowledge, based on the proper theoretical knowledge, in order to solve a practical task, and the ability to interpret the results in the frames of the use theory.

As an ability, competence is created during training (tab. 7.2). For instance the virtual laboratory is the environment for such a training in distance learning conditions. Quantitative and time-related measurement of a student's training results in the virtual laboratory, such as the speed and complexity of tasks solved within a given domain, allows for evaluating the degree of mastering (not just remembering) the proper theoretical knowledge, and the ability to use proper procedural knowledge, what might give evidence to the level of obtained competences. Complementing by the student the tasks given to him with new ones, which might be solved using the same theoretical knowledge, is the first step to develop his creativity.

Table 7.2 Evolution of the concept of competence

Competences based on abilities	Competences based on knowledge
– Few external stimulus (closed context) – Massive character of training (problems with individual adaptation) – Based on weakly formalized tacit knowledge – Problems with extending/adapting possessed competences to other domains	– Universality resulting from possessing fundamental knowledge (multi-domain) – Possibility to formalize basing on e.g. ontology engineering – Possibility of unambiguous presentation in the form of natural language, text, graphics – Necessity of constant complementation (awareness of the existence of gaps and lacks)

7.2.3 Cognitive Approach to Defining the Nature of Competence

Obtaining competences is most of all the effect of cognitive activity, therefore detailing the structure of competence requires a cognitive approach. Regarding computer systems, the approach to competences must be based on their cognitive nature, as it allows for gradual expansion of the intelligence of computers during performance of certain operations of information and knowledge processing.

Psychologists differentiate knowledge as such from the ability to use it, what is important in developing algorithms of knowledge processing and decision making in computer systems. Additionally, psychologists distinguish also the ability of specialists to gather knowledge from their ability to gather competences. In this context, the general intelligence is connected rather to creative productivity than the scope and depth of possessed knowledge. What is perceived as creative productivity is the ability of specialists to gather competences, what increases with time the scope of problems they study and the speed with which they solve these problems.

Simonton D.K., an American expert in the domain of organization analysis and scientific research management [31,32] states that competence is not derived only from the sum of a person's knowledge. It also depends, among others, on the ability to quickly analyze and structure indicators, characteristics, situations and tasks, what is the result of individual experience obtained through solving the researched problems. Results of experiments conducted by psychologists show that such ability bases not only on the innate nature of the person, but is also the result of properly organized training. According to this point of view, abilities obtained on the basis of individual experience are stored by the professional in the form of certain abridgements (pictograms) connected into a structure together with the mechanism for starting them. The number of abridgements together with the structure that integrates them increases with time, what leads to an increase in the speed of solving a problem. In computer science, the considered phenomenon can be described by a certain semantic network structure that contains a mechanism of scaling and granulating [1]. Psychologists connect creative productivity with the creation, in the memory of the specialist, of a system for typifying tasks and of algorithms for their successful solution on the basis of possessed knowledge is [31].

Simon D.K was for a long time the leader of the information approach in psychology. One of his projects concerned thinking models of a professional [30]. The obstacle in solving a problem/task is not only the insufficient level of knowledge, but also its excess. The dependence between knowledge and creative productivity can be described with a typical saturation curve consisting of three segments: the first one characterizes proportional dependence between the increase of knowledge and the increase of competence; the second one, in which the value of competence stabilizes at a certain level, attests to independence of creative productivity from continuing the process of knowledge acquisition, in other words, it attests to the saturation of the creativity potential of the person; the third segment indicates the existence of such a level of knowledge that is excess and causes decrease in creativity. The length of the segments has a wide diapason of values and depends on a set of psychological characteristics of the studied specialists. Allowing for wide interpretation, we can conclude that such a curve describes the border of openness of each system based on knowledge.

7.3   Ontology-Based Approach to Competence Modeling

Competences become increasingly important especially in such domains as knowledge management or distance learning. The essence of competence management is the need to adapt possessed competences to market requirements. The problem arises of referring the scope of competences to (or comparing it with) the job, technology and education market. In this section a formal model was proposed for evaluating the degree of adaptation of competences guaranteed by education offers to competences required on the job market after completion of the education process. The presented model can be generalized to comparing the scope of guaranteed competences to obtained competences.

7.3.1   Structure and Thematic Scope of Competence

On certain theoretical knowledge sets of different competences can be based [27]. In an educational situation, the scope of theoretical knowledge allows for formulating typical tasks, what makes it possible to create a hierarchical structure describing the thematic scope of a competence.

The first level of the hierarchy presented in figure 7.1. is the description of theoretical knowledge in the frames of the considered domain, with the use of a semantic network of basic concepts, defined as $N = \{S, R\}$, where $S = \{s_i\}$ - set of concepts, $R = \{r_q\}$ - set of relations between concepts. The entire network

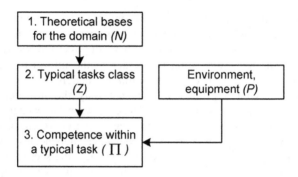

Fig. 7.1 Structure of the description of thematic scope of a typical task competence

can be described as a matrix $G = \left\| g_{i1,i2} \right\|,\ i_1,i_2 = 1,.....i *$,   where a matrix

elements $g_{i1,i2}$ is the description or relations between concepts (i_1,i_2).

At the next level of the considered structure one or more typical tasks $Z = \{z_j\},\ j = 1,..,j *$ are defined, each of which is connected to studying a certain type of process/system. From the point of view of methodology, the text of each tasks needs to be complemented with a referential model and a mathematical model.

From the point of view of the scope of theoretical knowledge, each typical task z_j can be represented by a subset of concepts used in the mathematical model, $S_j \subset S$. Subsets can intersect $S_{j1} \underset{j1 \ne j2}{\bigcap} S_{j2} \ne \varnothing$, and their union creates

the entire set $\bigcup_j S_j = S$.

The third level describes the scope of competences required to solve a typical task (tasks), i.e. it describes the scope of theoretical knowledge and the scope of procedural knowledge that are needed to realise the models described at the previous level.

The scope of procedural knowledge represents the technology of realising the mathematical model, which depends on the methodology and the software-hardware environment. In the description of procedural knowledge concepts from the subset $S_k \subset S$ must be used, and they have to be interpreted according to the terminology of the software-hardware environment P and the operations O^P, essential to solve task z_k in the considered environment.

The introduced notation enables describing the structure of competence c_k in the following way:

$$\Pi(c_k) = \{ S_k, S_k^P, O^P \},$$

where: S_k - set of concepts used in task z_k (theoretical knowledge), S_K^P - concepts from set S_k that require interpretation in the software-hardware environment (procedural knowledge connected to the theoretical knowledge), O^P - operations necessary to perform task z_k in environment P.

Obtaining competences in the frames of the structures presented in figure 7.1. is not a model of the real process, it is solely its computer metaphor. Applying such a metaphor gives the possibility to follow the results of the final, real cognitive process and to improve it by portioning and differentiating between theoretical and procedural knowledge.

7.3.2   Problem of Competence Definition Change in Time

Having to take the decision of entering the next level of education and beginning their studies young people need to face a huge challenge. With the increasing competition on the job market simply graduating does not guarantee a career. Studying becomes more and more common, thus the fact of having a university diploma is not something special anymore, it becomes a rather basic, standard element of the CV. The most important thing is what competences does the obtained education represent. It is thus very important to choose such an education offer that, through the offered abilities and knowledge, will assure the highest possible probability of finding a job.

Looking for a job is inherently connected to assessing the current state of the job market. Such assessment can be made through defining the requirement for certain qualifications (e.g. as understood within the European Qualification Framework as official confirmations of completed education) with regard to their number and scope. In the present times, however, such assessment is no longer sufficient. The current market state can be assessed with high probability only in the moment of beginning the education, the state of the market after the education is completed is weakly predictable, while changes occur too quickly for the prognosis based solely on information regarding currently needed qualifications to be enough to decide which studies will give a good chance of finding a job after they are completed.

Such dynamic changes on the job market are caused by quick outdating of knowledge and by development of technology in different domains of knowledge and life. The example presented in figure 7.2. depicts changes that may happen within one domain of the print media industry in regard to the included knowledge and technologies. This allows for concluding that qualification once obtained in a certain domain does not remain constantly up-to-date. What is more, the qualification obtained one year might mean something completely different than the qualification obtained a few years later. In such conditions it is important to replace predicting the requirement for job-positions on the basis of required qualifications with predicting on the basis of competence development.

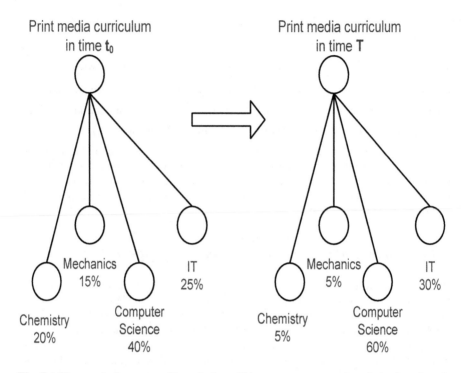

Fig. 7.2 Changes in the scope of knowledge within one competence domain in time (based on [27])

7.3.3 Formal Model of Comparing Guaranteed and Required Competences

The problem of choosing studies was always connected to the need of a different type of evaluation of alternative possibilities available on the education market. The situation roughly presented above allows for stating that the most important thing is to assess these possibilities from the point of view of the concordance of competences they offer with competences required on the market after the education is completed. Below authors present a formalized approach to this task [27].

1. $C^W = \{c_i^W\}$ - set of competences required on the labor market,

   $\quad G_i = \{V_i, y_i\}$ - graph of the required competence c_i^W,

   $V_i \in V$ - set of vertexes (concepts) of graph G_i,

   $y_i \in Y$ - set of arcs (relations between concepts) of graph G_i.

2. $D = \{d_k\}$ - types of technologies included in competences,

   $\quad H_k = \{w_k, x_k\}$ - graph describing competence d_k,

$w_k \in W$ - set of vertexes (concepts) of graph H_k,

$x_k \in X$ - set of arcs (relations between concepts) of graph H_k.

3. $G_k^i = G_i \bigcap H_k$ - sub-graph of technology d_k, used in competence c_i^w,

$\bigcup\limits_{k=1}^{k*} G_k^i = G_i$ - conjunction of all graphs G_k^i creates the competence graph G_i,

$G_{k1}^i \bigcap G_{k2}^i \neq \emptyset$ – intersection of the subgraphs is not empty.

4. $\mu_k^i = \dfrac{\left| G_k^i \right|}{\left| G_i \right|}$ - participation of technology d_k in competence c_i^w,

$0 \le \mu_k^i \le 1$.

5. $F\ (G_k^i, t)$ - *characteristic of dynamic of subgraph* G_k^i,

$F\ (G_k^i, t)$ is a function that characterizes variability of graph G_k^i,

There are three possible types of changes in graph G_k^i:

    1) Appearance of new vertexes (concepts) in graph G_k^i,

    2) Elimination of vertexes (concepts) from graph G_k^i,

    3) Modification (change of the relation between concepts) in graph G_k^i.

Let us introduce the following parameters of changes:

λ_{ik}^1, R_{ik}^1 - intensity and distribution of 1st type changes,

λ_{ik}^2, R_{ik}^2 - intensity and distribution of 2nd type changes,

λ_{ik}^3, R_{ik}^3 - intensity and distribution of 3rd type changes,

Then the variability of graph G_k^i can be characterized by the following functional:

$$F\ (G_k^i, t) = F(\ \lambda_{ik}^1, R_{ik}^1,\ \lambda_{ik}^2, R_{ik}^2,\ \lambda_{ik}^3, R_{ik}^3\ )$$

6. $T_N = [t_0, T]$ - *cycle of mastering competence* c_i^w, *(learning cycle)*,

T_N is the cycle in which we observe changes in required competences.

7. $G_k^i(t_0)$ - *subgraph of technology* d_k *in competence* c_i^w *at time* t_0,

$G_k^i(t_0)$ is the state of graph G_k^i at t_0, which can be characterised by the vector $\overline{M}_k^i(t_0)$

$\overline{M}^i(t_0) = \{\mu_1^i(t_0), \mu_2^i(t_0),\ldots\ldots\mu_{k*}^i(t_0)\}$ - participation of each technology d_k in competence c_i^W at time t_0.

8. $G_k^i(T)$ - *subgraph of technology* d_k *in competence* c_i^W *at time* T ,

$G_k^i(T)$ is the state of graph G_k^i at T , t_0, which can be characterized by the vector $\overline{M}_k^i(T)$,

$\overline{M}^i(T) = \{\mu_1^i(T), \mu_2^i(T),\ldots\ldots\mu_{k*}^i(T)\}$ - participation of each technology d_k in compete1nce c_i^W at time T.

9. $[P]^i$ - *matrix of transition between states of graph* G^i *at times* t_0 *and* T ,

$[P]^i = \|p_{kl}\|^i$, where $k = 1,2,\ldots k *$ - set of technologies in competence c_i^W ,

$\quad\quad\quad\quad\quad l = 1,2,\ldots l *$ - degree of changes (+ or -) in technology during time interval (t_0,T) .

An element of the matrix, p_{kl}, is the probability of changes in technology d_k in degree l after the interval of time (t_0,T)

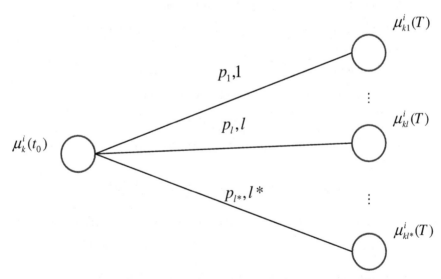

Fig. 7.3 Matrix of transition between the states of graph G^i at times t_0 and T

10. $\overline{M}^i(T) = \{\mu_1^i(T), \mu_2^i(T), \ldots\ldots\mu_{k*}^i(T)\}$ - vector of participation of technology in competence c_i^W at time T,

Vector $\overline{M}^i(T)$ is connected to vector $\overline{M}^i(t_0)$ by the transition matrix $[P]^i$ with the following product (fig. 7.3.):

$$\overline{M}^i(t_0) \times [p_{ik}^T] = \overline{M}^i(T)$$

Participation of k technology in competence c_i^W at time T can be defined on the basis of Bayesian (weighted average) probability

$$\mu_k^i(T) = \sum_{l=1}^{l*} \mu_{kl}^i(T) p_{kl} = \sum_{l=1}^{l*} \mu_{ik}^i(t_0) \cdot l \cdot p_{kl}$$

11. $G_i^W(T)$ - *summary graph of required competence c_i^W at time T* ,

$$G_i^W(T) = \bigcup_{k=1}^{k*} G_k^i(T), \text{ where}$$

$G_k^i(T)$ - subgraph of technology d_k in competence c_i^W at time T .

12. $S(c_i) = \{s_j\}$ - *set of education offers covering the required competence c_i^W* ,

$c_j^\Gamma \in C^\Gamma$ - competence guaranteed by offer s_j ,

$C^\Gamma = \{c_j^\Gamma\}$ - full set of competences guaranteed by each offer in the scope of the required competence c_i^W .

13. $R_j^\Gamma = \{u_j, z_j\}$ - *graph of guaranteed competence $c_j^\Gamma \in C^\Gamma$* ,

$u_j \in U$ - set of vertexes (concepts) of graph R_j ,

$z_j \in Z$ - set of arcs (relations between concepts) of graph R_j .

14. $Q(R_j^\Gamma, G_i^W)$ - *matching degree of guaranteed competence $R_j^\Gamma(t_0)$ from offer s_i to required competence $G_i^W(T)$ at time T* ,

$Q(R_j^\Gamma, G_i^W) = R_j^\Gamma(t_0) \bigcap G_i^W(T)$ intersection of the graph of required competence $G_i^W(T)$ at time T with the graph of guaranteed competence $R_j^\Gamma(t_0)$ at time t_0 .

15. $F(R_j^\Gamma) = |Q(R_j^\Gamma, G_i^W)|$ - *function of determining the degree to which the guaranteed competence $R_j^\Gamma(t_0)$ matches the required competence $G_i^W(T)$* .

Formulation of the task of comparing guaranteed and required competences

Input data

1. Set of required competences $C^W = \{c_i^W\}$ and technologies included in these competences $D = \{d_k\}$,

2. Ontological graph of the chosen competence $c_i^W \in C_W$, $G_i = \{v_i, y_i\}$ and graphs of technologies included in competence c_i^W, $H_k = \{w_k, x_k\}$,

3. Competence c_i^W mastering cycle, $T_N = [t_0, T]$,

4. Set of education offers containing the required competence c_i^W,

$$S(c_i^W) = \{s_j\},$$

5. Graphs of guaranteed competences of each offer,

$$R(c_j^\Gamma) = \{u_j, z_j\}, \ c_j^\Gamma \in C^\Gamma,$$

6. Subgraph of each technology G_k^i, used in competence c_i^W, $G_k^i = G_i \bigcap H_k$ and participation of technology in competence c_i^W,

$$\mu_k^i = \frac{|G_k^i|}{|G_i|}, \ 0 \le \mu_k^i \le 1,$$

7. Characteristics of dynamics of each subgraph G_k^i (intensity and distribution of variations in the graph)

$$F(G_k^i, t) = F(\lambda_k^i, R_k^i),$$

8. States of subgraphs $G_k^i, k = 1,...k*$ (participation of each technology in competence c_i^W) at times t_0 (beginning of the learning cycle) and T (end of the learning cycle)

$$\overline{M}^i(t_0) = \{\mu_1^i(t_0), \mu_2^i(t_0), ...\mu_{k*}^i(t_0)\}$$

$$\overline{M}^i(T) = \{\mu_1^i(T), \mu_2^i(T), ...\mu_{k*}^i(T)\}$$

With given:
  - input data (1-8) describing the state of the competence market,
  - for a chosen competence c_i^W

The following has to be defined:

-    set of education offers $s_j \in S$,
-    that meet the criterion

$$Q(c_j^\Gamma, c_i^W) = R(c_j^\Gamma, t_0) \bigcap G(c_i^W, T)$$

$$\left| Q(c_j^\Gamma, c_i^W) \right| = \underset{s_j \in S}{Max} - \delta$$

where
δ represents the space of possible deviations from the maximum according to user's preferences

The presented assessment model can be used in a computer system aimed at supporting a user in choosing an education offer. Such a system, using the predictions of competence requirements on the market, considering the possibilities and the predispositions as well as preferences of a user, would help define a set of education offers that should be taken into account while making the choice.

7.4 Competence Set Approach to Competence Modelling

There is still a lack of well developed formal models allowing for creation of quantitative methods for competence analysis. One of the most advanced ideas of this type is the approach called competence sets (tab. 7.3.), introduced for the first time by Yu and Zhang [36,38]. It provides methods for competence set analysis and its quantitative models thus giving a solid background for development of competence management systems that can be easily implemented and applied in many domains.

Table 7.3 Main characteristics of the competence sets method (source [26])

Characteristics	Description
Inventors	Yu and Zhang
Background literature	[36,38]
Main scientific results:	competence set analysis approach, competence set expansion process, competence set expansion costs and expansion rewards, optimal expansion process.
Decision theory application	consumer decision problem, generating learning sequences for decision makers, model of decision problems in fuzzy environments,
Knowledge management application	consumer decision problem, generating learning sequences for decision makers, model of decision problems in fuzzy environments.

7.4.1 Competence Set Theory

In the initial phase of the development of the competence set theory the competence was seen as a classical set [36] containing knowledge, skills and information necessary to solve a problem. Further studies of the nature of competence lead to the decision that due to its continuous character it is not sufficient to assess the presence of a competence in binary terms (either entirely present or not at all). Therefore, it was proposed to present competence as a fuzzy set, defined in the following way [39]:

$$A = \{(x, \mu_A(x)) | x \in X\}$$

where: $\mu_A(x)$ is the function assessing membership of an element x in relation to set A by mapping X into membership space [0; 1], $\mu_A : X \rightarrow [0; 1]$.

For each person P and each event E, in regard to which the competence g is assessed, it is possible to define the competence as a function of this person or event that will reflect the strength of the considered competence: $g^\alpha(P)$ or $g^\alpha(E)$, where α is the fuzzy competence strength and is defined in the following way: $\alpha : \{P \text{ or } E\} \rightarrow [0; 1]$

Above all this, yet another concept was created. Scientific research regarding methods of knowledge acquisition shows that this process is more successful if a proper order is maintained [2]. In other words, a certain portion of knowledge is easier to master if basic knowledge connected to it is acquired before. In the competence set theory this basic knowledge is regarded as background competence. The pace with which new competences can be learned results from the level (strength) of the obtained background competences and on how closely the background competence relates to the considered competence.

Therefore, if (for any person P) competence g_1 with strength α_1 is the only background competence of competence g_2, then the potential, referred to as background-strength, of obtaining g_2 equals $\beta_2 = \alpha_1 \cdot r_{21}$, where $0 < r_{21} = r(g_1, g_2) \leq 1$ reflects the background relation between these two competences. The stronger the background relation and strength of the background competence, the easier it is to obtain a new competence. Competences g_1 and g_2 are independent if $r_{12} = 0$ and $r_{21} = 0$, in other words neither is the background competence of the other. If a competence has several background competences then the learning process progresses according to the principle of maximal support [17].

Additionally, the critical level γ of the background strength β needed to obtain competence g is defined. The value of the critical level γ depends on the competence, it is denoted by $\gamma(g)$ and is from range [0; 1], so that if $\beta \geq \gamma$ then cost $c = 0$, otherwise $c > 0$. A person willing to obtain competence g has to possess such background strength of this competence that $\beta \geq \gamma(g)$. If the condition is fulfilled, competence g is called the person's skill competence. Otherwise, g is called the person's non-skill competence. The set of all skill competences is denoted by $Sk(P)$, while that of non-skill competences is denoted by $NSk(P)$ [34].

According to Wang [34] competence $g^{\alpha(E),\beta(E)}$, required for person P to solve problem $E,$ is called person P's:

- type (1) competence if $\beta(P) \geq \beta(E)$ and $\alpha(P) \geq \alpha(E)$,
- type (2) competence if $\beta(P) \geq \beta(E)$ and $\alpha(P) < \alpha(E)$,
- type (3) competence if $\beta(P) < \beta(E)$.

The set of all competences necessary to solve problem E is defined as $Tr^{\psi,\omega}(E)$, where vector ψ defines required strengths of the competences in Tr and vector ω defines their necessary background strengths.

7.4.2 Quantitative Assessment of Competence

Basing on the presented competence set theory, it is possible to perform quantitative assessment of competences. For this, it is essential to first properly identify the sets of acquired (possessed) and required competences: Sk and Tr respectively. In the first case, universally accepted competence standards, different best practice databases or taxonomies can be used to help with the identification. Identifying the acquired competences requires focusing only on detecting competences from set Tr in the person being tested. Figure 7.4. presents the process of quantitative competence assessment in more detail.

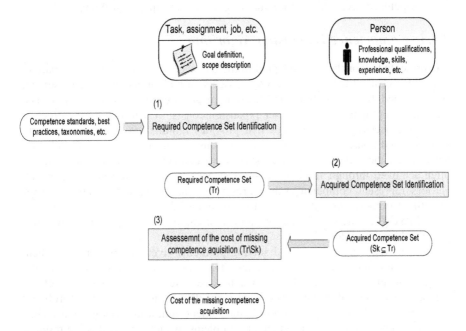

Fig. 7.4 Assessment of competence expansion cost (source [26])

The entire process of assessing the cost of acquiring missing competences consists of three stages:

1. Identification of the set of competences required to realize a task. First of all, basing on the description of the goal and scope of the task, all competences required to successfully perform this task need to be identified. In the simplest case, if the task classifies to the typical, often realized kind of tasks, it is possible to find for it some standardized competences using one of the existing standards or norms. In case the task is not typical and there are no official competence standards applicable for it, the abilities required to realize this task can be identified through expert analysis. The expert performing this analysis can use his own experience as well as different sources of knowledge regarding the domain of the task, such as all kinds of textbooks, articles, knowledge compendia, etc.

2. Identification of the set of competences possessed by the student. On the basis of the set of competences required to realize the task, as identified at the first stage, it is possibly to identify these competences in the student. This can be achieved through analysis of previously realized projects, scientific experiments, publications, reports, etc.

3. Assessing the cost of acquiring missing competences required to realize the task. On this stage the Quantitative analysis of the cost of expanding competences through comparison of the set of competences possessed by the student with the set of competences required to realize the task is preformed. This analysis can be performed basing on any of the mathematical models of competences available in literature (e.g. a model proposed in Wang 1998).

7.4.3 Competence Expansion Algorithm

The presented algorithm was adapted to the Distance Learning Network situation. In [17] the cost estimation algorithm and decision-making model were presented, as developed for curriculum modification in an educational organization. The methods of optimal competence set expansion consist of determining the order of obtaining successive competences that provides minimal cost [4,9]. In learning system conditions, the functionality of this algorithm focuses mainly on determining the order of obtaining successive competences by students. This is often achieved through finding the shortest path in an oriented graph, in which vertices represent competences and arcs represent the relations between them [34,36]. Competences that need to be obtained are defined by the set $Tr^{\psi,\omega}(E)/Sk^x(P)$, where x is the vector of strengths of competences currently possessed by the student P.

Students realizing projects and solving problems given and presented to them possess many individual characteristics. The most important of which is the different level of base knowledge, which is reflected in the different levels of competences. Very often this basic level is insufficient to solve a given problem. Developing a solution requires training the student in new knowledge areas. Thus it is very important to assign students to problems in such a way, that the training is effective. We can identify the following goals of training:

1. Strengthening already possessed knowledge, repetition of already known material. The goal is to choose such problems, that solving them would require using already mastered knowledge for the purpose of solidifying it.
2. Exploring knowledge, unlimited discovering of new knowledge, e.g. in the frames of scientific procedures connected to a task. Solving the given problem requires looking for solutions in different areas and domains, regardless of the cost and resources.
3. Teaching/learning with constraint on resources, standard mode of education, where we base on resources existing in the system, e.g. servers, teachers etc. Using each of these resources generates specified cost.

New knowledge should be available with low cost and based on knowledge already possessed by the student. The truth is that each student can solve each problem, but the cost of such an approach is quite substantial. The cost of expanding competences is connected to the time and resources we need to put into achieving the required degree of knowledge in the student. For example, costs can be generated through buying access to a given portion of knowledge (Learning Object) [40] or teacher's working time.

If we gave up the proposed analysis of competence expansion cost, problems (in the form of projects or other activities) would be assigned randomly. A student with high level of competence in a given area could be assigned a project that does not bring anything into his basic knowledge or requires too much, while the goal of learning assumes otherwise. From the point of view of cognitive processes, the best problem to solve is the one for which the scope of required competences is directly connected to the student's base knowledge [40].

The following assumptions have been made [17]:

Input data:

$P = \{p_i \mid i = 1,2,...,i^*\}$ - set of students in learning system

$HD^{\alpha,\beta} = \{g_k^{\alpha_k,\beta_k} \mid k = 1,2,...,\infty\}$ - space of all competences related to teaching activities

R – relation matrix of competences in HD

$Sk^x(p_i) = \{g_j^{x_j} \mid g_j^{x_j} \in HD^{\alpha,\beta}\}$ - competence set of i-th student

E – problems to be solved (in the form of projects or other activities)

$Tr^{\psi,\omega}(E) = \{g_l^{\psi_l,\omega_l} \mid l = 1,2,...,l^*\}$ - set of competences (courses) necessary to solve the problems

$c\left(Sk^x(p_i), g_l^{\psi_l,\omega_l}\right)$ - cost function of obtaining competence g_l by student P_i

C_o – assumed maximum cost of solving the problems

Control parameters:

$U = \|u_{il}\|$ - allocation matrix

$$u_{il} = \begin{cases} 1, \text{student } p_i \text{ competence set } g_l \\ 0, \text{otherwise} \end{cases}$$

Assuming, that a single course can be allocated only to a single student:

$$\sum_{l=1}^{l^*} u_{il} = 1$$

Criterion:

1) $\Phi = \sum_{l=1}^{l^*} \left[u_{il} \cdot \delta_i \cdot c\left(Sk^x(P_i), g_l^{\psi_l, \omega_l} \right) \right] = \min_U < C_R < C_o$ - training goal: repetition, value C_R tends to 0.

2) $\Phi = \sum_{l=1}^{l^*} \left[u_{il} \cdot \delta_i \cdot c\left(Sk^x(P_i), g_l^{\psi_l, \omega_l} \right) \right] = \max_U \leq C_o$ - training goal: knowledge exploration

3) $\Phi = \sum_{l=1}^{l^*} \left[u_{il} \cdot \delta_i \cdot c\left(Sk^x(P_i), g_l^{\psi_l, \omega_l} \right) \right] \min_U \leq C_S < C_o$ - training goal: learning with constraints on resources (limiting resources C_S)

If all necessary competences are covered by competences of all students, we can say that the students' group is able to deal with the problem. Thus, it is necessary to present a new definition of competence types, one that takes into consideration joint competences of a group. Hence, we can classify group competences as follows:

1) $g^{\alpha,\beta}(E)$ is called a type (1) group competence if

$$\exists_{i=1}^n (\beta(P_i) \geq \beta(E) \wedge \alpha(P_i) \geq \alpha(E)),$$

2) $g^{\alpha,\beta}(E)$ is a type (2) group competence if it is not a type (1) group competence and $\exists_{i=1}^n (\beta(P_i) \geq \beta(E) \wedge \alpha(P_i) < \alpha(E))$,

3) $g^{\alpha,\beta}(E)$ is called a type (3) group competence if $\forall_{i=1}^n \beta(P_i) < \beta(E)$.

In figure 7.5. the heuristic algorithm of group competence expansion (Group Competence Expansion Algorithm - GCEA) was presented. As can be seen, two main phases were identified in the expansion process [17]:

1) Preliminary expansion phase. The preliminary expansion phase contains activities that are performed to increase strengths of competences already possessed by the group, when these strengths are too low in regard to the given problem E. This corresponds to transforming all type (2) into type (1) competences. The preliminary expansion phase contains steps 1 – 4 of the algorithm in figure 7.5.

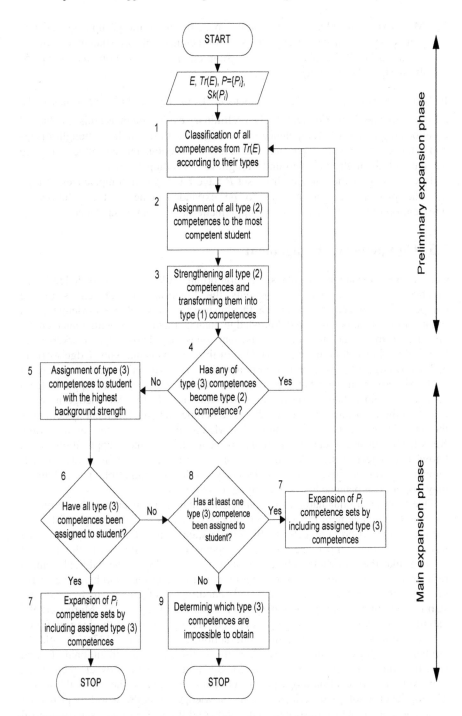

Fig. 7.5 Group competence expansion algorithm (based on [17])

2) Main expansion phase. In the main expansion phase the group acquires new competences that have not been skill competences of any student within the group so far. It this phase, type (3) group competences are obtained (steps 5-10 of algorithm GCEA).

In both expansion phases every competence from the set $Tr^{\psi,\omega}(E)$ has to be allocated to exactly one student. In the preliminary expansion phase this is achieved by allocating a type (2) competence to the student with the highest strength of this competence. In the main expansion phase, a type (3) competence is allocated to the student with the highest background strength for this competence.

In case none of the students from set P have background competences of type (3), the process of learning has to be expanded. One of the solutions is intervention of the teacher, who can supplement the missing competence.

7.5 Competence Management

Competence management plays an important role in a knowledge-based organization [29]. Figure 7.6. presents the functions and objectives of the competence management system in the business context of a knowledge-based organization. Competence and knowledge management deals with management objects, from which the most important ones are knowledge workers and knowledge objects (competence set). One of the reasons is that knowledge workers and knowledge objects directly influence the organization's intellectual capital [21]. Moreover, a well-built competence management system supports team cooperation and collaboration in order to create an effective knowledge network based on the associated workflow [25]. The competence management system uses the knowledge network to achieve knowledge diffusion in the organization. Knowledge diffusion can be defined as adaptations and applications of a knowledge object in the knowledge network environment [3]. It is created only if knowledge workers involved are sufficiently similar or compatible, i.e. as long as they display competence levels that are not too far apart [7].

Due to their similar nature, the competence management process bases on the knowledge management process. Both processes use the same strategy (Tab. 7.4) with regard to different objects. To fully understand competence management, one has to define the competence set. The competence set has its root in the habitual domain, like the knowledge object has its root in ontology. The habitual domain reflects a person's unique set of behavioral patterns resulting from his or her ways of thinking, judging, responding, and handling problems, which gradually stabilized within a certain boundary over a period of time [35]. Moreover, the habitual domain incorporates the person's collection of ways of thinking, judging, etc., accompanied by its formation, interaction, and dynamics [37]. The classic Gruber's ontology definition described the ontology as an intentional, formal specification of concepts in a given domain and relations between them [11]. However, the top-level ontology [10] is a better equivalent of the habitual domain. The top-level ontology describes general knowledge about the world by providing basic concepts about time, space, events and conditions. These terms are universal and can be used to create specific domain ontologies.

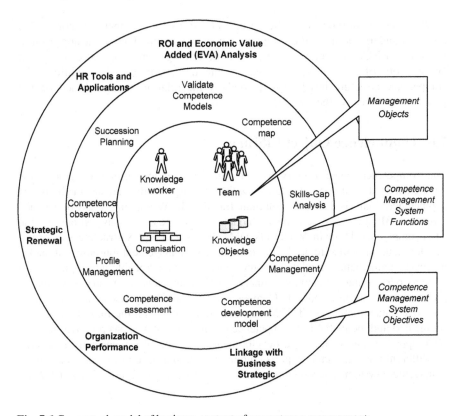

Fig. 7.6 Conceptual model of business context of competence management

Table 7.4 The similarity of knowledge and competence management (based on [24])

	Knowledge management	Competence management
Objects	Knowledge object	Competence set
Codification strategy (person-to-document approach) Based on the externalization process (tacit to explicit) [20].	Encodes and stores knowledge in online databases and various repositories where it can be easily used by any knowledge worker.	Encodes and stores the competence in online databases and various repositories where it can be easily used by any knowledge worker.
Personalization strategy (person-to-person approach) Based on the socialization process (tacit to tacit) [20].	Creates, uses and shares knowledge peer-to-peer, supported by appropriate communication facilities.	Creates, uses and shares knowledge peer-to-peer, based on the competence profile, supported by appropriate communication facilities.

The most important scientific result related to the competence concept is common understanding of the complex competence nature. One of the approaches, from the TENCompetence project [28], is describing competence structure as an aggregation of: proficiency level, context, description. In addition, the competence structure has been extended by the relation strength based on the fuzzy competence set theory [26].

7.6 Competence Object Library

The Competence Object Library presented in [26] is an approach to integrated new competence description standards with existing mathematical methods of competence analysis into one common framework. Within the TENCompetence project [33] the most complete model of competence was proposed: the TENCompetence Domain Model (TCDM). It has a complex structure supporting development of software tools for competence management purposes, it lacks, however, mathematical models of competence that would enable development of a system performing quantitative analysis of competence. Authors present a proposition for using a reusable Competence Object Library (COL) that integrates objective data structures of TCDM into one framework with the fuzzy competence set method and the method of competence expansion cost analysis proposed in [17]. Such approach allows for rapid development of competence management systems for different purposes, and supports communication interface inside the competence management system. The structure of COL was presented in figure 7.7. and in [26].

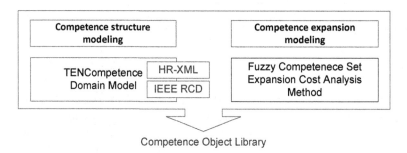

Fig. 7.7 Concept of the Competence Object Library (source [26])

In COL elements of competence's description have been connected with a method for competence processing. The main element is a competence connected to a proficiency level and a context. Such understood competence is a part of competency, which can be treated as the habitual domain. The COL competence element stores information about each competence with related strength attributes in the form of competence set.

The structure of COL was modeled with UML as a UML Class Diagram. UML was chosen due to the fact that it is the most common and recognizable notation

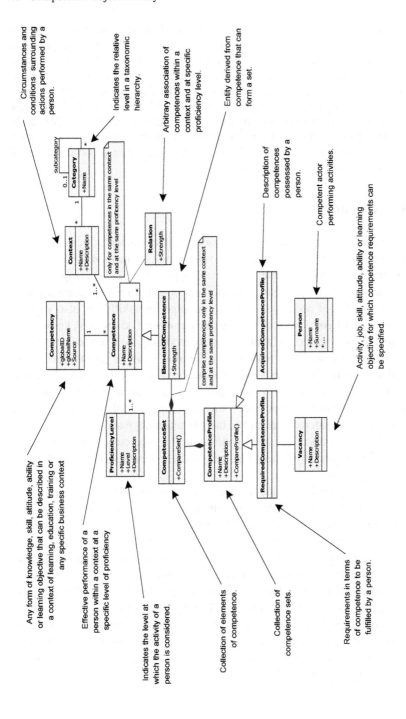

Fig. 7.8 UML Class Diagram for Competence Object Library (based on [24])

among software engineering professionals. The UML Class Diagram depicted in figure 7.8. combines the two background approaches: TENCompetence Domain Model and fuzzy competence sets. TCDM was used as the basis for class structure modeling, and the original structure was extended with classes representing notions from the fuzzy competence sets theory, what allowed adapting COL to its analytical methods operating on sets, relations and graphs defined by these relations. The algorithm of computing the competence expansion cost [26] can be called with two methods: CompetenceSet.CompareSet() and Competence-Profile.CompareProfile(). These methods return a real value proportional to the difference between two competence sets/profiles.

7.7 Conclusions

In this chapter different aspects of the concept of competence were presented. Investigation of this matter allowed for development of a method for formalizing this concept on the basis of the ontological approach and the competence set theory apparatus. The scope of the definition of competences is not constrained by ISO as „demonstrated ability to apply knowledge and skills", it also includes interpretation of different types of knowledge: know what, know how, know why.

The first of the proposed approaches to modeling competences treats the structure of competence as a resource of different types of knowledge. In the frames of this approach, the issue of competence change in time was discussed. On the basis of the presented assumptions a formal model of comparing guaranteed and required competences was developed. In the proposed model, competences are considered in regard to the job, technology and education market. The model assesses the degree of conformity of competences guaranteed by education offers with competences required on the job market after completion of the education process.

The second approach uses the competence set theory. The proposed method works at the level of knowledge and employs competence sets as knowledge representation models in educational organizations. Basing on the group competences expansion algorithm it is possible to asses and optimise the cost of expanding competences The algorithm focuses on the cost of student competence expansion caused by the knowledge development process.

The Competence Object Library is a tool for rapid development of applications for competence management and analysis. The main idea of COL is to provide HR and e-Learning professionals with reusable library allowing developing applications for quantitative competence analysis. Additionally, the interface to open competence exchange standards like IEEE RCD and HR-XML creates new possibilities for web-centric HR and e-Learning systems.

References

1. Anderson, J.R.: Cognitive Psychology and Its Implications, 5th edn. Worth Publishing, New York (2000)
2. Ausubel, D.P., Novak, J.D., Hanesian, H.: Educational Psychology: A Cognitive View, 2nd edn. Holt, Rinehart and Winston, New York (1978)
3. Chen, C., Hicks, D.: Tracing knowledge diffusion. Scientometrics 59(2), 199–211 (2004)
4. Chen, T.Y.: Using competence sets to analyze the consumer decision problem. European Journal of Operational Research 128, 98–118 (2001)
5. Council of Europe: Key competencies for Europe. Report of the Symposium in Berne (March 27-30, 1996); Council of Europe, Strasbourg (1997)
6. Doninger, N.A., Kosson, D.S.: Interpersonal construct systems among psychopaths. Personality and Individual Differences 30(8), 1263–1281 (2000)
7. Ehrhardt, G., Marsili, M., Vega-Redondo, F.: Diffusion and growth in an evolving network. International Journal of Game Theory 34(3), 383–397 (2006)
8. EQF: European Qualifications Framework for Lifelong Learning, http://ec.europa.eu/education/lifelong-learning-policy/doc44_en.htm
9. Feng, J.W., Yu, P.L.: Minimum Spanning Table and Optimal Expansion of Competence Set. Journal of Optimization Theory and Applications 99(3), 655–679 (1998)
10. Gomez-Perez, A., Fernandez-Lopez, M., Corcho, O.: Ontological Engineering. Springer, Heidelberg (2004)
11. Gruber, T.R.: A translation approach to portable ontologies. Knowledge Acquisition 5(2), 199–220 (1993)
12. HR-XML: HR-XML Consortium (2006), http://www.hr-xml.org/
13. IEEE 1484.20.1/draft - draft standard for Reusable Competency Definitions (RCD), http://ieeeltsc.wordpress.com/working-groups/competencies/
14. ICOPER - Interoperable Content for Performance in a Competency-driven Society. An eContentplus Best Practice Network (2008-2011), http://www.icoper.org/
15. IMS RDCEO: Reusable Definition of Competency or Educational Objective (RDCEO), http://www.imsglobal.org/competencies
16. ISO 24763/draft: Conceptual Reference Model for Competencies and Related Objects (2009)
17. Kusztina, E., Zaikin, O., Różewski, P., Małachowski, B.: Cost estimation algorithm and decision-making model for curriculum modification in educational organization. European Journal of Operational Research 197(2), 752–763 (2009)
18. Lahti, R.K.: Identifying and integrating individual level and organizational level core competencies. Journal of Business and Psychology 14(1), 59–75 (1999)
19. Mansfield, R.S.: Building competency models: Approaches for HR professionals. Human Resource Management 35(1), 7–18 (1996)
20. Marwick, A.D.: Knowledge management technology. IBM Systems Journal 40(4), 814–830 (2001)
21. Nemetz, M.: A Meta-Model for Intellectual Capital Reporting. In: Reimer, U., Karagiannis, D. (eds.) PAKM 2006. LNCS (LNAI), vol. 4333, pp. 213–223. Springer, Heidelberg (2006)

22. Perrenoud, P.: Construire des compétences dès l'école. Pratiques et enjeux pédagogiques. ESF éditeur, Paris (1997)

23. Romainville, M.: L'irrésistible ascension du terme compétence en éducation, Enjeux, (37/38) (1996)

24. Różewski, P., Małachowski, B.: Competence-based architecture for knowledge logistics in project-oriented organization. In: O'Shea, J., Nguyen, N.T., Crockett, K., Howlett, R.J., Jain, L.C. (eds.) KES-AMSTA 2011. LNCS, vol. 6682, pp. 630–639. Springer, Heidelberg (2011)

25. Różewski, P.: A Method of Social Collaboration and Knowledge Sharing Acceleration for e-learning System: the Distance Learning Network Scenario. In: Bi, Y., Williams, M.-A. (eds.) KSEM 2010. LNCS, vol. 6291, pp. 148–159. Springer, Heidelberg (2010)

26. Różewski, P., Małachowski, B.: Competence Management In Knowledge-Based Organisation: Case Study Based On Higher Education Organisation. In: Karagiannis, D., Jin, Z. (eds.) KSEM 2009. LNCS, vol. 5914, pp. 358–369. Springer, Heidelberg (2009)

27. Różewski, P., Kusztina, E., Sikora, K.: Formal approach to required and guarantee competence comparison. In: Owsiński, J., Nahorski, Z., Szapiro, T. (eds.) Operational and Systems Research, Exit, Warszawa, Poland, pp. 433–440 (2008) (in Polish)

28. Sampson, D., Fytros, D.: Competence Models in Technology-Enhanced Competence-Based Learning. In: Adelsberger, H.H., Kinshuk, P.J.M., Sampson, D. (eds.) Handbook on Information Technologies for Education and Training, 2nd edn., pp. 155–177. Springer, Heidelberg (2008)

29. Sanchez, R.: Understanding competence-based management: Identifying and managing five modes of competence. Journal of Business Research 57(5), 518–532 (2004)

30. Simon, H.A.: The information-processing theory of mind. American Psychologist 50(7), 507–508 (1995)

31. Simonton, D.K.: Scientific creativity as constrained stochastic behavior: The integration of product, process, and person perspectives. Psychological Bulletin 129, 475–494 (2003)

32. Simonton, D.K.: Age and outstanding achievement: What do we know after a century of research? Psychological Bulletin 104, 251–267 (1998)

33. TENCompetence - Building the European Network for Lifelong Competence Development (2005–2009), EU IST-TEL project, http://www.tencompetence.org/

34. Wang, H.-F., Wang, C.H.: Modelling of optimal expansion of a fuzzy competence set. International Transactions in Operational Research 5(5), 413–424 (1998)

35. Yu, P.L., Lai, T.C.: Knowledge Management, Habitual Domains, and Innovation Dynamics. In: Shi, Y., Xu, W., Chen, Z. (eds.) CASDMKM 2004. LNCS (LNAI), vol. 3327, pp. 11–21. Springer, Heidelberg (2005)

36. Yu, P.L., Zhang, D.: A foundation for competence set analysis. Mathematical Social Sciences 20, 251–299 (1990)

37. Yu, P.L.: Habitual Domains. Operations Research 39(6), 869–876 (1991)

38. Yu, P.L., Zhang, D.: Optimal expansion of competence sets and decision support. Information Systems and Operational Research 30(2), 68–85 (1992)

39. Zadeh, L.A.: Fuzzy sets. Information and Control 8, 338–353 (1965)

40. Zaikin, O., Kusztina, E., Różewski, P.: Model and algorithm of the conceptual scheme formation for knowledge domain in distance learning. European Journal of Operational Research 175(3), 1379–1399 (2006)

Part III

Application of Open Learning Systems

Chapter 8
The Concept of a Virtual Laboratory as the Space for Competencies Transfer

8.1 Introduction

The problem of knowledge management has become increasingly important in the mainstream of contemporary science. From the information technology standpoint, the predominant research problems are knowledge storing and knowledge sharing, determination of methods dealing with knowledge expansion and refinement along with adapting knowledge structures to individual expectations (knowledge personalization). Furthermore, the scientific areas related to cognitive science are primarily focused on identifying knowledge structures stored in human mind and further development of the methods pertaining to knowledge engineering. However, a problem such as knowledge digitalization is largely distinct from the aforementioned, as it is oriented towards processing of knowledge resources such as the Semantic Web.

In the recent years, the research problem of knowledge management has been extended over the aspect of human competencies. The competences concept represents the amount of knowledge acquired by a student (individual) in the course of learning process. Denoting one's knowledge by competencies gives us an opportunity to formulate a systemic description of a given problem domain that gravitates towards actual qualifications of a given individual. Moreover, using the competencies formula allows to define a problem domain as a collection of subsequent classes that represent an increasing degree of specialization (progressing from novice to expert). Additionally, applying the competencies abstraction renders a precise domain description possible by specification of the associated actions and abilities in forms of comprehensible textual information.

From a pragmatic point of view, the goal and purpose of any educational organization is to prepare competent and qualified personnel that will serve the needs of the economy and society. The foundations for the competencies and qualifications are the acquired knowledge and overall experience. Laboratory classes are of significant importance in the process of providing theory and accumulating practical skills by learning individuals. Due to Internet possibilities the virtual laboratory becomes the most important instance of laboratory class. Inclusion of laboratory work in the learning cycle on the one hand requires preparation of a studying individual. On the other hand, it necessitates a properly

P. Różewski et al.: Intelligent Open Learning Systems, ISRL 22, pp. 179–211.
springerlink.com   © Springer-Verlag Berlin Heidelberg 2011

defined laboratory environment. What's more, the student preparation process in the laboratory is comprised of supplying and subsequently verifying knowledge that is essential for completing the given task. In conventional learning, maintaining both the control and even complementing the student's knowledge in ongoing fashion are executed by a laboratory tutor. Contrastingly, the role of the tutor in the virtual laboratory is replaced by a suitably designed software component (the other alternatives would be the knowledge model in form of repository, the expert system or the ontological domain model). The issue of designing the type of software is discussed in [14].

8.2 Virtual Laboratory Concept

The virtual laboratory is becoming an increasingly important element in the knowledge transfer space realized in the contemporary university and polytechnic. That situation is influenced by several factors. Firstly, as indicated by [3], the tutoring of engineers using online systems is more effective due to the cost element and facilitated access to the tutoring resources. The ongoing development of computer hardware and network infrastructure extend the range and capacities of the virtual laboratory, for example by offering access to either virtual instrumentation and machinery [10], or remotely controlled experiments conducted within real performance of the devices [26]. Secondly, certain scientific areas have engaged in more extensive use of information technologies and dedicated software leading to reducing the distance to the virtual learning environment. The areas such as chemistry [9], or geography [22] have been gradually provided with computer systems, and that fact facilitates the design of the corresponding elements of the virtual laboratory. Lastly, the human factor can be introduced. The over decade-long research activities dealing with human-to-computer interaction discussed by [6], have allowed to adapt the learning individual's work space to his cognitive and psychomotor requirements. The result of those is the sequence of predefined work places arranged both ergonomically [27] and technically [8].

The concept of virtual laboratory has been acknowledged and widely discussed in literature. [7] has defined virtual laboratory as a heterogenic distributed environment that allows group of agents (e.g. students or researchers) to carry out joint project activities. In [16,24], the concept of virtual laboratory has been extended with three mutually complimentary types of laboratory. The evolution of the concept was initiated by unsophisticated systems simulating the activity of a given artefact in a constrained manipulation environment, such as operation of an injection moulding machine. The next stage is constituted by open systems based on application of sophisticated mathematical models. The last generation of virtual laboratory that is the subject in question in this chapter are the objects dealing with domain knowledge transfer that include the cognitive factors and knowledge of the recipient.

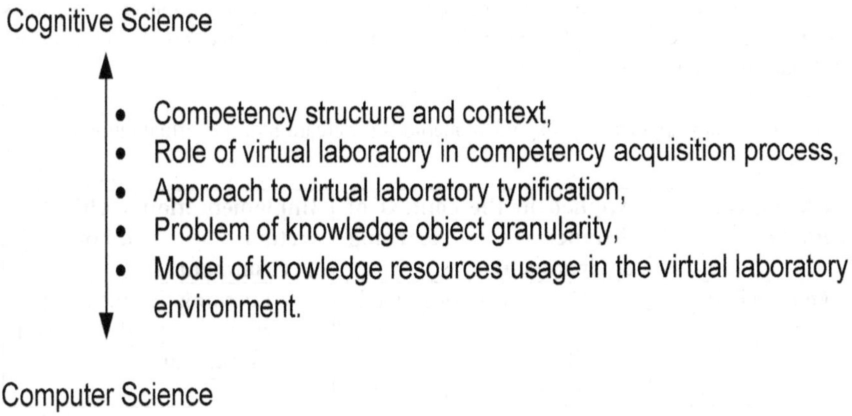

Cognitive Science

- Competency structure and context,
- Role of virtual laboratory in competency acquisition process,
- Approach to virtual laboratory typification,
- Problem of knowledge object granularity,
- Model of knowledge resources usage in the virtual laboratory environment.

Computer Science

Fig. 8.1 Different aspects of Virtual Laboratory research problem

The goal of a laboratory operating on the level of domain knowledge is extending student's abilities (competencies) to formulate a problem and identifying the tools applicable to resolve it within conceptual limits of the given domain (figure 8.1.). The laboratory allows to develop analytical skills using theoretical knowledge [24]. The student's knowledge increase occurs at solving problems that are characteristic of given domain and related to different domain-specific objects and different processes selected from the domain. Furthermore, application of methods of artificial intelligence that originate from cognitive science, such as the structure of a term (concept) and problem solving method is also characteristic in this context. The open space of the virtual laboratory is defined by a learning event that is aimed at creating optimal conditions for knowledge acquisition, for example through application of the knowledge already acquired by the student to formulate computerized learning metaphors.

8.2.1   The Goals and Methodical Foundations for Conducting Laboratory Classes

The topic and purpose of laboratory classes should be reflected in proper identification of proportions between theoretical and procedural knowledge transferred to learning individual during laboratory classes. Therefore, identification of the foundations for conducting the identification of proportions between the types of knowledge has been a subject of research activities performed by pedagogical and teaching methodology experts [5]. Table 8.1 introduces a synthesis of key methods for choosing adequate computerized learning metaphors by information technology experts. The implementation of the methods in the context of virtual laboratory is only possible by means of applying the adequate computerized learning metaphor. It distinguishes from modeling applications by lack of requirements for maintaining the relationship between the goal, the subject and other elements pertaining to the teaching process. Consequently, it enables interpreting those elements accordingly in the framework of appropriate information system.

Table 8.1 The key applicable pedagogical methods for creation of the virtual laboratory

Pedagogical context	Applied in the context of distance learning environment	Implementation within the virtual laboratory environment
Behaviorism	– Focused on observable student's performance characteristics. – Trial&error approach	– Learning event bound with predetermined checkpoints.
Cognitivism	– Learning process is aimed at providing pre-identified cognitive structures (objects) for learner's mind. – Identification of knowledge objects transferred to the student.	– Method of knowledge models development [29]. – Algorithms for creation and identification of knowledge objects [17].
Constructivism	– Concentrated on observing student's interaction with environment. – Providing the student with access to resources that offer active interface.	– Dedicated educational corporate-style portals, such as Moodle. – Providing with dedicated educational tools, e.g. simulation software, statistical analysis software.

8.2.2 Fundamental Components of the Virtual Laboratory

As a resource for the didactical process, virtual laboratory environments should be composed of three main elements:

- networked computer environment comprising of the software and the hardware enabling a learning individual to connect remotely to the virtual laboratory's contents.
- specialized software as the means for performing task-specific operations, i.e. manipulating the virtual objects, performing modeling, simulation, etc.
- didactical materials repository supporting the process of self-conducted learning by studying individuals.

Each of the aforementioned components is characterized with distinct structure. Hence, planning its utilization will require applying different management models that were discussed in [16]. The specialized software and didactical materials repository components comprise the virtual laboratory contents and as such it can be identified as a tuple:

$$K = \{D,OP,R\}, \text{ where:}$$

D - a domain that can be described with determined granularity (research area, subject, topic, facet)
OP - the framework used for modeling, or performing simulation of a given problem
R - didactic materials with determined didactic goals with collection of typical examples and tests and ontological or reference models, etc.

In conventional teaching, a student performing laboratory course tasks assisted by laboratory teacher, or tutor utilizes the appropriate didactic materials (instructions, course outline, list of course specific tasks) and equipment. On account of that, the following question can be posed: how should the didactic materials be changed and what should the physical hardware be superseded with speaking from the perspective of the virtual laboratory environment. The answer to this question predominantly depends on the course's topic, defined teaching goals and hardware- and software-related constraints allowing establishment of the virtual learning space. For that reason, preparation of the laboratory environment is mainly concerned with creating an appropriately designed information system. Thus, the virtual laboratory is an information system that is formed of the following components:

a)   computerized environment fully taking over the role of any hardware equipment (multimedia-based space for interaction).
b)   model of profiled domain knowledge accommodating the course topics
c)   repository storing tasks and their solutions
d)   scenario for conducting the course
e)   mechanism for measuring the effectiveness of addressing a given task
f)   means for remotely and interactively accessing to all of the aforementioned elements

Lack of physical presence of a teacher, or tutor in the virtual laboratory should be compensated with providing the a - f elements to the leaning individuals and increasing their engagement . Moreover, maintaining access for the student in terms of the elements of b and c provides him an opportunity for addressing the deficient object of knowledge on the fly. In that context and from the information science standpoint, the main goal for the virtual laboratory that operates within the characteristics of the information system is identifying a mechanism for flexible adaptation of didactical materials to the required proportion between theoretical and procedural knowledge while maintaining the determined course didactic goals and curriculum.

8.2.3 Typing of the Virtual Laboratory

Specificity of constructing the components of the virtual laboratory is related to the identified didactic goals and curricula [16]. From the pedagogical point of view, a broad range of laboratory course goals can be identified that can be enveloped by contrasting cases: from acquiring skills and confidence in running devices to capacity for formulating problems, conducting analysis and reasoning. That is why foundation of the virtual laboratory requires resolving a variety of problems among which the predominant one is establishing the relationship between information and cognitive sciences aspects. Further analysis of that problem allows us to identify the following three scenarios and corresponding laboratory types [16,24]:

A. The didactic goal: acquiring the skills for controlling unsophisticated devices. In that context, the curriculum is aimed at relaying to students the qualifications required for conducting repeatable operations with the devices in precisely determined work places. In that sense, the complete virtual laboratory system is similar to a computer game such as the Tetris insofar as the laboratory software includes rudimentary animated models allowing for their iterated manipulation according to a given use case. That facilitates teaching straightforward problems, i.e. assembly manufacturing operations.

B. The didactic goal: acquiring the skills for controlling all of the hardware, devices and runtime procedures that require formalized complex mathematical modeling with elements of visualization. The curriculum is aimed at relaying to students the qualifications necessary for controlling sophisticated devices under conditions of uncertainty. In that sense, the complete virtual laboratory system is a simulator acting between the device and the student in various use cases. The software used for that type of laboratory is an intelligent system encompassing a set of parameterized mathematical, simulation and animation models targeted at creating a virtual reality for the student. Under those circumstances the learning individual performs a predefined role with accordance to qualifications required for becoming a pilot, or a driver. A good example of such software is the concept of virtual gauging and monitoring equipment presented in [21], or the ASIMIL system introduced by [20].

C. The didactic goal: acquiring the skills for conducting an analysis of a given phenomenon, situation, or process, problem statement, reasoning based on theoretical knowledge, or finding a method for representing a new concept. The curriculum is aimed at influencing and shaping creativity, systemic and abstract thinking. Such qualifications are applicable not only to managerial staff positions, but also to research personnel. The required software should facilitate interactive knowledge sharing among teaching participants during laboratory-, or project-related teaching activities and maintain the connection between theoretical and procedural knowledge during activities such as conducting research experiments and interpreting collected data and information. The complete virtual laboratory system in that situation is composed of a variety of models allowing for representing a given phenomenon on different levels of abstraction with a model of appropriate type, e.g. reference, mathematical, or conceptual. It is therefore mandatory to maintain a variety of distance learning standards (such as the SCORM), learning management systems (such as the Moodle software) with an applicable teaching method. The student operating in the laboratory of this type, performs a role of researcher conducting an experiment within the framework of intelligent virtual environment.

In each of the aforementioned scenarios it is possible to perform a more detailed classification applying different classification criteria, for example such as the used software, or the formed simulation models.

Table 8.2 The main characteristics of virtual laboratories

Types	Main characteristics	Didactic goal	Scientific problems
1	- using artifacts - simulating simple operations	Acquiring appropriate habits	Human-computer communication, Interface developing issue
2	- controlling and operating an object - the object is represented by a mathematical model - working in real time mode - creating virtual reality environment	Learning procedures and understanding functioning rules of an object	The effect of presence and immerse, modeling complex objects, real time systems
3	- leading a discussion (deliberation) and analysis according to a specified methodology - the object is represented by an abstract model - working out and using ontologies	Acquiring the ability to formulate a problem and to find a mean and method of solving it	Modeling knowledge representation and use, intelligent knowledge-exchange environment

8.2.4 *Example of Virtual Laboratory Working on Knowledge Level*

An example of the knowledge virtual laboratory was described in [14] and presented on figures 8.2 and 8.3. The Simulation Web Portal was developed in order to introduce the principals of computer simulation to the students. A simple online simulator is a good opportunity for students without any additional preparations to make their first step into the topic of computer simulation. The Simulation Web Portal consist of four main modules/sections (figure 8.2): a list of simulation principles, a collection of online tests, the online simulation module and personal records management. Within the laboratory framework, the students participate in modeling queuing systems with a help of the Arena package and the SIMAN language (figure 8.3.) [30]. A user of Online Simulator cannot develop its own simulation model because that functionality is limited only to running a simulation of predefined models.

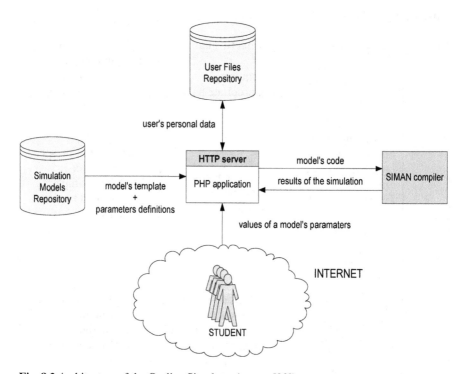

Fig. 8.2 Architecture of the On-line Simulator (source [30])

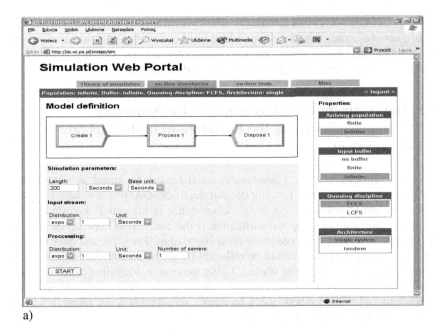

a)

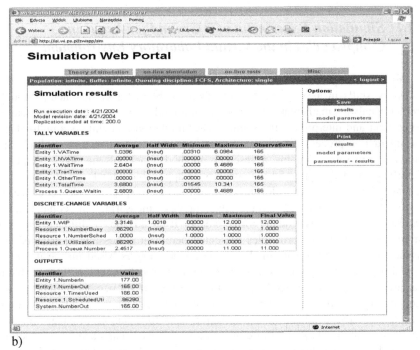

b)

Fig. 8.3 The Simulation Web Portal. a) The Model Definition view. b) The Simulation Results view (source [30]).

Other example of virtual laboratory is a CyberMath environment [13]. In this case, a network space was developed based on using the OpenGL standard, where the teacher and the students are represented by avatars and learn advanced mathematical concepts through manipulation and interaction.

8.3 Development of an Environment for Acquiring Competencies Using the Foundations of the Virtual Laboratory

A general didactic goal of a laboratory' activities can be stated as establishing personal competencies and forming desired qualifications. It is a widely accepted belief that the foundation for a certain competence is theoretical knowledge and the foundation for obtaining a qualification is the connection between theoretical knowledge with an appropriate procedural knowledge. The topic and didactic goal for a laboratory' activities should be reflected by proportions between those types of knowledge presented to the student during the course. Identification of rules for establishing the proportions is a research goal for pedagogical experts and didactic methodologists. That is how the laboratory' activity plays an integral role in acquiring various skills and knowledge.

8.3.1 Simulation Experiment as a Ground for Acquiring Competencies in the Virtual Laboratory Environment

A simulation experiment can prove to be a useful ground for providing the necessary skills in a very broad sense: from problem (task) statement, through problem analysis, mathematical and simulation modeling to conducting a research experiment and interpreting the results. The simulation modeling is a multi-stage and interactive process. Dividing it into stages is a consequence of qualities the simulation process contains a research method. Each of the stages requires identification of the type and the contents of applicable knowledge and that fact becomes the foundation for development of a problem-specific tasks sequence. It is therefore reasonable to claim that precise measuring of competencies acquisition process is equally effective as it is in the traditional learning environment. Using the principles presented herein, it is possible to establish an instance of a laboratory of simulation. The laboratory is based on the use of a typical model repository classified according to Kendall's notation. That problem has already been analyzed in the context of a procedure for personal competencies acquisition. The content and order of student-oriented tasks is a result of a widely accepted routine dealing with preparation and formulation of stages in the simulation experiment [2,4,23]. Let us examine the particular stages of conducting the simulation procedure in the perspective of the tasks provided to the learning individuals.

Problem analysis

At this stage both the limit of the problem domain and exogenous limitations (of time, financial, ecological, ergonomic, or other nature) specific to the researched phenomenon, system, or object should be identified. Moreover, it is necessary to provide conclusion if simulation as the research method is applicable. The means for satisfactory problem statement is a reference model of the phenomenon under investigation. Construction of such model is key for the entire simulation modeling process and it directly impacts the process' stages, especially the requirements for input data, length of a simulation model development process and its validity and the types of experiments targeting such model. At this stage, the didactic material should contain a collection of reference models among which only a few should qualify for further research using the simulation-based experiment. In such context, the least sophisticated student-targeted problem is identification of a problem that can be solved using the given mathematical apparatus.

Problem statement

Problem statement that serves as a roadmap and task formulation for all of the parties involved in creation of the simulation model, prevents a situation of failing to follow the defined research objective. Using the results conducted beforehand, the following are to be determined:
- the research objectives
- the input and output elements of the analyzed system
- the analyzed system's structure
- assumptions concerning both the unknown and the uncertain system elements
- model formulation trade-offs that need to be applied with accordance to the modelled real-time system

One of the most important features of the correctly conducted problem statement process is its advanced abstraction. It implicates that the task should be independent of any simulation environment and technique.

Speaking from the perspective of a manufacturing process simulation laboratory, it is feasible to identify typical research problems and perform their classification. That facilitates the upcoming analysis and implementation phases and also leads to certain degree of simulation workflow automation using the given reference models. The didactical material at that stage should contain so called problem links that can be identified as a pair: an identified problem - a set of multi-element task formulations.

For example, let's review the following problem of creating a corporate network for the purpose of distributed intangible production system. The following tasks can be formulated: a) identification of the corporate network's structure and configuration, b) optimization of of workplaces efficiency, c) identification of the network channel's bandwidth. Among these, only tasks (b) and (c) are likely to require a formulation of simulation experiment. The task (a), however, is carried out using mathematical model without resorting to simulation.

The student is expected to correctly identify the tasks that will require simulation experiment formulation.

Formulation of the mathematical model

Formulation of the mathematical model for the given process is usually extremely abstract. Its main advantage when compared to the simulation model is computing speed and decreased costs of development. Usually, its main downside is related to its limited computing precision in comparison with the simulation model. More importantly, use of the simulation model is effective in performing verification of the mathematical one. The most straightforward way of performing verification is comparing the results form both models or applying methods of statistical analysis. At this stage it is advisable to use mathematical models based on traditional methods which feature greatly refined classification schemes, such as those found in queuing theory (i.e. Kendall's notation) [31]. Kendall's notation is a universal research instrument. In the virtual laboratory it is applied not only for performing classification of mathematica models, but also to classify and identify other simulation models. The task aimed at students is carrying out the analysis of the mathematical model and interpretation of its parameters in relation to the observed real-time system.

Defining input data

Defining input data. There is a very big correlation between the structure of the model and requirements regarding the input data used for the experiment. Together with the change in complexity of the model, the requirement for input data, gathered and processed in a proper way, also changes. For example, if the referential model is defined as a one-phase process, the given parameter – operation duration – is interpreted as one number, whereas when the process is defined as a multi-phase one, it is necessary to define the vector of duration of operation performance in each phase of the process. It is the student's task to track how the changes in the referential model influence the definition of input data. In more complicated tasks it is also necessary to track how the change in the nature and scope of the parameters influences the structure of the mathematical model – what will have influence on the choice of verification method.

Algorithmisation and programming of the simulation model

After properly formulating the research problem and developing the referential model, on the basis of which the task assumptions were built, we can progress to the stage of implementing the simulation model. As was already mentioned, properly defined task assumptions are independent of any simulation environment. This gives freedom in choosing the package in which the building of the simulation model and conducting of simulation experiments will start. The choice can be made between such solutions as: GPSS/H, Arena, AutoMod, CSIM, Extend, Micro Saint, ProModel, Deneb/QUEST or WITNESS.

For the needs of the virtual simulation laboratory, in the considered examples the Arena package of Rockwell Software Inc., which is a graphical frontend for the simulation language SIMAN, was used. The main argument for using this package is the fact that SIMAN gives great possibilities for storing and reusing simulation models in a repository developed especially for this purpose. Typical models can be stored in the repository in a parameterized form, due to what their use is brought down to solely downloading a proper model from the repository and defining parameter values. In such case, the process of conducting a simulation experiment is shortened by the stage of simulation model building. The role of the student at this stage is to skilfully use the SIMAN repository.

Model verification and validation

Both of the stages constitute the most sophisticated and significant challenges to the model analyst (designer). It should be mentioned that validation should not considered as a merely separate collection of procedures emerging sequentially after each model development phase has been completed, but it should become an integral part of the model development process. Validation performed on the fly during the model development process is considered to be more effective than the one committed once the development has been completed and is therefore of greater usefulness in detecting model defects. The end result of the validation process is a binary one and it determines whether the model firmly reflects the analyzed real objects. The effect of validation is usually achieved through running an iterative comparison analysis conducted between the simulation model and the actual performance of the real system under investigation. The process is repeated as long as its precision is found acceptable.

The purpose of verification is providing an answer if the model has been developed correctly and if so to what degree. The student's task is to address the following questions: have all of the experiment assumptions been satisfactorily implemented? Have all of the input parameters and the model's logical been correctly reflected in the analyzed model? Can the simulation model be used to verify its mathematical counterpart?

Parameters optimization

Optimization of mathematical model parameters is of very complex nature and can be executed with use of various methods of optimization [19]. The key find here is to identify the optimization method that is most suitable to the given research problem. Optimization of the model parameters can be aided with use of specialized software that contains several optimization algorithms implemented and a mechanism for their selection. An example of such software are Autostat (the AutoMod suite), OptQuest, SimRunner (the ProModel suite), Output Analyzer (the Arena software) [2]. The main goal or the student is to formulate a criterium of optimization applicable to a given example. The problem of using optimization software tools is not a part of the problem discourse.

Result interpretation

Interpretation of results is an analysis of output data generated during either the experiment, or a series of simulation experiments. Since the greatest number of cases is characterized with stochastic nature of the output data, their analysis necessitates application of stochastic methods. This stage is concluded with a definite answer addressing the initial research goal identified during the problem statement phase. The form specific to the results interpretation process may be distinctive and relies both on the researched problem's nature and the recipient of the processed results. In addition, it is important to use an adequate concept model. If the research object is a queuing system network, the result data should be presented in the terms characteristic to the queuing systems theory and Kendall's notation [31]. The student is asked to present the conclusive results using the terms of the reference model in place.

As shown in the description above, the simulation experiment can be used as a sort of battleground individually-controlled learning as well as a platform for collective project development. The condition for using the simulation experiment in such situation is to examine coherence between the terms specific to given stage of the simulation model development.

8.3.2   Methodology for Developing the Simulation Model

Development of the simulation model requires initial problem conceptualization and specification of the given problem using terms of a relevant domain. Problem conceptualization may usually be represented by inconsistent structure and is often implemented using unsophisticated description formed using natural language. The inherent diversity of concepts in that case and limited precision of specification of the problem statement create major problems at the stage of modeling. The most complex issue is maintaining accuracy in the process of transformation of concepts of the analyzed domain to terms and functions applied in a given simulation tool, i.e. to a certain portion of procedural knowledge [29].

Simulation software usually includes libraries of predefined templates for modeling different problem domains. Unfortunately, in most cases they are fairly limited in scope and applicability and their are designed to respond to select, or most typical problems. On the other hand, they effectively handle entire classes of problems within a scope of given mathematical topic, i.e. fundamental (theoretical) knowledge that is represented by well defined structure and concept taxonomy. In such case it is possible to identify and formulate an official analytical model. Moreover, a mathematical apparatus and formal notation are available for precise analytical task formulation. We can refer to the example of Kendall's notation serving the queuing system theory [31].

Therefore, the stage of problem statement using the concepts of universally known mathematical engine and its relevant notation can be seen as less complex in performing transformation of the domain concepts to formulas used in the given simulation software than in the example of direct transformation to the terms of a chosen domain characterized by limited precision of its concepts. That is why the previously mentioned stages of simulation experiment statement conducted in the environment of virtual laboratory should be complemented by several iterations of mapping of the verbal problem representation onto the simulation model (figure 8.4). In that context, documents such as the 2000 Mathematics Subject Classification in which mathematical taxonomy is introduced play a pivotal role in facilitating the process of identification of adequate mathematical engine upon formalizing the problem domain.

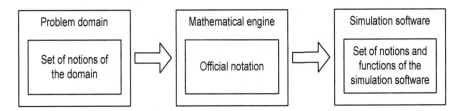

Fig. 8.4 The stages of mapping the conceptual model in the process of simulation experiment

As it has been shown on figure 8.4., the simulation experiment statement is exemplified by use of the concepts that are specific to the given simulation software. The concepts form a language for defining the problem and requirements essential for conducting the simulation experiment. The preparation of the concepts is not possible without completing the stage of forming the assumptions about the specific problem using the concepts of the domain that the analyzed problem is pertinent to. The simulation experiment statement performed in such approach is difficult to be translated to the language (terms and formulas) of the simulation software. Hence, additional modeling process is performed using a mathematical apparatus that is adequate to the given problem. Using that measure significantly simplifies the simulation process of the mathematical model using the given simulation software.

It is vital for the process of formulation of initial assumptions about the researched problem to identify a criterion for performing analysis of the model (figure 8.5.). Once the criterion has been determined, its examination is required in order to identify its characteristics (continues, or discrete), or to find other significant parameters. The problem containing properly defined criterion can be subject to modeling using the procedure of mapping the concepts of fundamental knowledge onto procedural knowledge. The mapping can be carried out through identification of direct equivalents between distinct conceptual models [17].

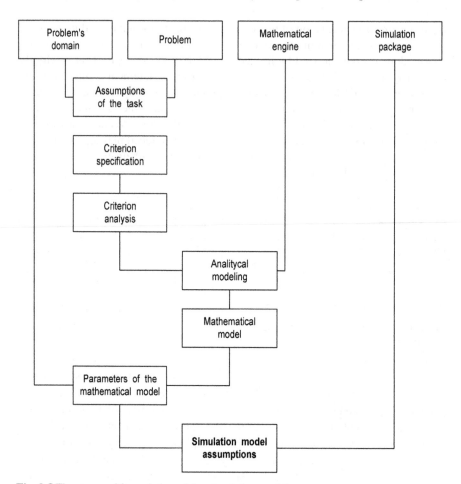

Fig. 8.5 The stages of formulation of the simulation model

8.3.3   The Algorithm for Determining Simulation Experiment Settings

The algorithm for determining the simulation experiment settings will be discussed using an example of a research problem defined using queuing systems theory. Using a problem statement and the mathematical apparatus of queuing systems theory, an elementary event occurring at the input to the system is to be determined. Naturally, the event is considered the most significant component of the model. In addition, all the parameters involved in the event input stream are to be determined, and their average intensity (i.e. a number of events on a given time frame and their characteristic: continuous, or discrete). If the input event stream is of stochastic nature, a probability distribution function of time intervals between

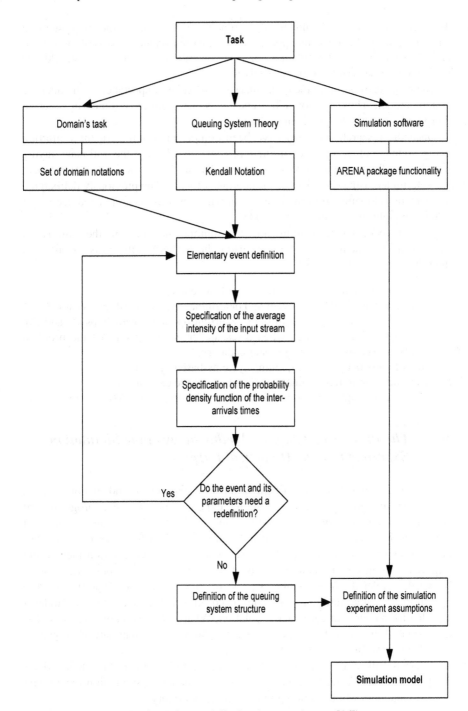

Fig. 8.6 Algorithm of the simulation model's development (source [16])

the incoming events should be determined. In a queuing theory system, a numerous types of events can appear, and for that reason the process of the system identification and determination of the operational parameters should be performed for each of the type respectively.

The process of defining the elementary events is key importance to the queuing system model. Thus, it should be repeated some number of times in order to thoroughly investigate the given problem.

Once the research activities have been completed resulting in the definition being sound and acceptable, it then possible to identify the rationale laying foundations for the simulation experiment. The assumptions for the experiment are determined using the ARENA simulation software. The previously deliberated mathematical-to-simulation model transformation process is conducted according to the principles of concept mapping [18].

To conclude, the algorithm for determining rationale of the simulation experiment for queuing system theory using the ARENA software is contained in the following steps (figure 8.6.):

1. Given a problem domain, perform a task statement.
2. Determine an elementary event and determine parameters of the event flow.
3. Refine the event definition through recurrent operation with steps (1) and (2). The process is terminated when a researcher is convinced that the problem can be solved using queuing systems theory.
4. Identify a structure of the queuing systems-based system.
5. Determine an initial outline (specification) of the experiment.
6. Create a corresponding simulation model using the ARENA software.

8.3.4 The Process of Adapting Methodology of the Simulation Experiment to the Didactical Purposes

The previously debated stages of the simulation experiment and its composition are commonly applied practices among experts and computer programrs of simulation modeling. In order to effectively teach those activities carried out within the Internet-based environment, a suitable system for teaching computer-aided simulation and an extensive domain knowledge representation method are required. The most significant elements of such system are the repositories storing knowledge related to particular stages of the simulation experiment. The repositories enable multiple and automated knowledge relay to the students enrolled for a course in computer simulation. What's more, a repository offered at each stage of the simulation experiment algorithm is an inherent part of a common knowledge repository in a given teaching system.

Figure 8.7 illustrates a process of defining a simulation experiment. It reflects the procedure that a learning individual upon conducting simulation experiment needs to maintain while operating in the virtual laboratory.

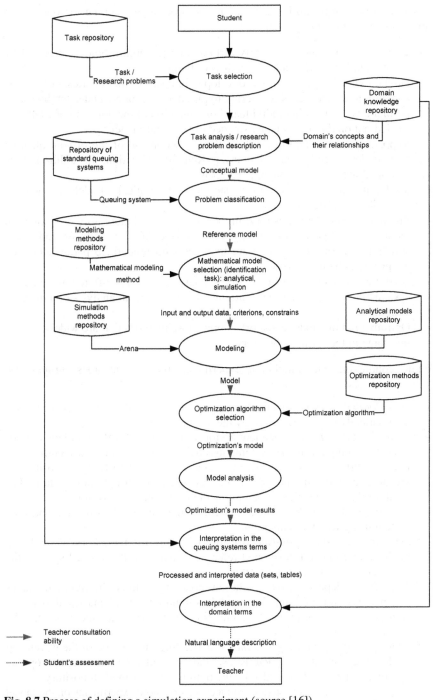

Fig. 8.7 Process of defining a simulation experiment (source [16])

The following repositories are used in the process of teaching to conduct the simulation experiment (see figure 8.7):

- Task repository: stores a variety of tasks and research problems accessed by the student prepared to address and solve them.
- Domain knowledge repository: developed in forms of dictionaries, or lexicons containing all the necessary terms required for accurate understanding of a problem by the learning individual. Formalization of the domain knowledge is carried out on the basis on an extended ontological model.
- Repository of standard queuing systems: contains all of the official queuing system designations classified using Kendall's Notation. In addition, it is equipped with concept mapping patterns.
- Modeling methods repository: stores mathematical modeling methods available to the student.
- Simulation methods repository: in the case of virtual laboratory of simulation modeling the repository is strictly related to the ARENA software and stores simulation methods (i.e. software modules with predefined parameters) that are provided by the software.
- Analytical models repository: stores analytic methods offered by queuing systems theory that are used to arrange and facilitate an analysis of a queuing system.
- Optimization methods repository: stores those methods that are widely used at this stage in the virtual laboratory.

While conducting the simulation experiment, the student goes through the following stages:

- *Downloaded task* – student download a task to solve or is assigned one automatically by the system or by the teacher.
- *Analysis of the task* – the student performs an analysis of the problem, defines inputs and outputs of the system, structure of the system, and identifies the most important parameters. At this stage a domain knowledge repository is useful. It provides definitions of concepts used for formulating the task. The result of this stage is the referential model of the analyzed system.
- *Classifying the problem* – if it is possible, the student classifies the analyzed system using Kendall's notation.
- *Defining the model type* (identification) – using the repository of modeling methods and the self-developed referential model of the studied system, the student chooses a modeling method.
- *Modelling* – depending on the chosen modeling method, different techniques can be used for building the model: simulation ones, analytical ones, or combined ones. Appropriate simulation and analytical methods can be downloaded from the repository.
- *Choosing and optimization method* – depending on the type of the developed model and on its characteristics, the student chooses an appropriate optimization method, provided by the optimization methods repository.
- *Studying the model* – the student conducts one or more experiments on the model in order to obtain results necessary to answer the stated research question.

- *Interpretation of results in the terminology of queuing systems* – output data obtained as a result of the experiments is interpreted. First of all, the interpretation should be performed in the terms of queuing systems, and the data should undergo statistical analysis.
- *Interpretation of results with the use of concepts from the problem domain* – giving an answer to the task in natural language, using the same concepts that were used to formulate it.

The above steps of the procedure are performed in a distance way in the environment of the virtual simulation laboratory, however, after each of them is finished it is possible to consult the teacher which is supervising the learning process.

8.4 The Method for Competence Verification in the Virtual Laboratory Environment

The advantages of competence-based knowledge management have been used by the European Union to create a coherent system for formal representation of knowledge relayed at consecutive cycles of higher education process. Within the scope of the European Higher Education Area, every student is fully capable of shaping his own educational advancement relying on competencies designation offered and delivered through courses available at universities [15]. The representation format of the individually acquired competencies makes identification and assessment of student's capabilities attainable to different university, or an employee. Currently, the problem is under thorough examination by the bologna Working Group on Qualification Frameworks. The concept of the Bologna Process assumes creating an all-European system of qualifications (competencies) [15].

Unfortunately, the current advancement of areas as a domain of research does not enable to use the effective methods for competencies evaluation of a given individual. The approved tools for knowledge assessment mainly relate to the learning-teaching process. A good example of such tool can be found, such as the exponential learning curve, or other teaching-related models described in [12]. As a consequence, a new need for competencies evaluation tools and mechanisms is required that will also operate in the context of the virtual laboratory., or a distance learning course based in the methodology of knowledge objects methodology [29].

8.4.1 Management of Competences Acquisition Process in the Context of the Virtual Laboratory

The structure of competence can be described in the following way:

$$\Pi(c_k) = \{S_k, S_k^P, O^P\},$$

where: S_k - set of concepts used in task z_k (theoretical knowledge), S_K^p - concepts from set S_k requiring interpretation in software-hardware environment (procedural knowledge joined with theoretical knowledge), O^P - operations required for performing task z_k in environment P. The proposed approach enables creating a knowledge repository based on a knowledge model represented by the triple <theoretical knowledge (what), procedural knowledge (how), project knowledge (where)> according to the algorithm (figure 8.8.).

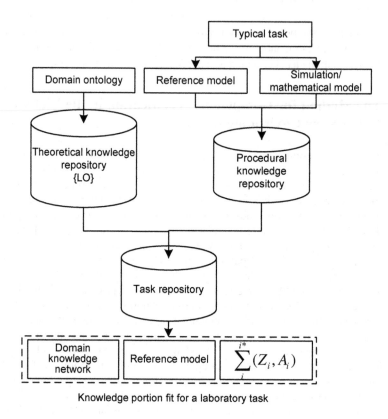

Knowledge portion fit for a laboratory task

Fig. 8.8 The laboratory tasks repository collection algorithm

8.4.2   *The Procedure of Competencies Acquisition*

The procedure of competencies acquisition shown on the figure 8.9. was originally introduced by [16,24]. The purpose of the procedure is to:
–   provide training of the competencies acquisition process to the students, given: a level of theoretical knowledge, kinds of required competencies, and a length of time of a single training session;
–   collect statistical data enabling specification of the management models related to personalization of learning process given the teaching environment.

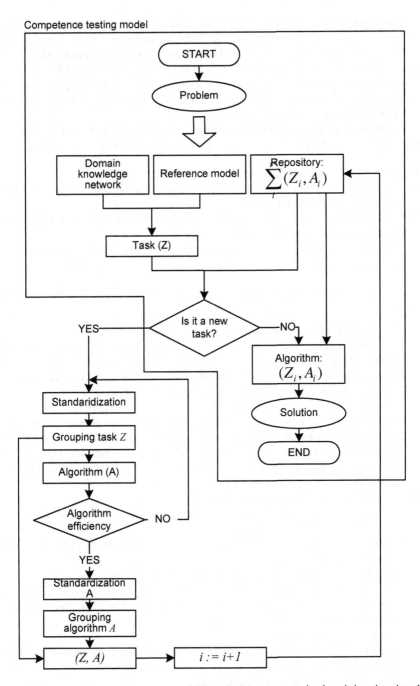

Fig. 8.9 Procedure of competence acquisition during an organized training in virtual laboratory environment (source [24])

Using the virtual laboratory repository, a student is provided with a collection of tuples: a portion of domain specification - typical task - typical solution and with a corresponding test task. The domain specification is formulated using the extended ontological model identified in [17]. The test task should be interpreted using the terms specified using the proposed tuple. The approach presented herein enforces the student's capacity to structure the accumulated knowledge and merge it with results of his own experience.

Let us discuss the procedure according to the material presented by [24]. The input data of the procedure are identified in the following way:

1. Domain: the subject/topic of study.
2. A model of theoretical domain knowledge (built on the basis of the methodology proposed in [29].
3. A reference model, enabling the use/development of taxonomy of the problems (tasks) studied during the training.
4. A repository of task solutions: the repository may be based on solutions proposed in [14].

The competence acquisition procedure by [16] is achieved with the following steps:

1. Research problem analysis.
   Deciding if the problem pertains to the domain specified in the task. That enables the processes of domain interpretation and specification using the terminology of the specified knowledge model with respect to taxonomy. In other words, the step is focused on formulating the input task.
2. Analysis and systematization of learning individual's experience.
   Comparing the contents of the input task with the tasks already placed in the repository. As a result, a solution pattern is established that has to be followed in order to solve the given problem. We choose either development of a solution-appropriate algorithm or use an algorithm already defined in the repository.
3. Standardization of the input problem.
   Preparing the input task token in the repository language (e.g. in the form of an XML document).
4. Accumulating the input-task token in the operational memory of the current training session.
5. Developing an individual algorithm for problem solving.
   The algorithm may be described with pseudo-code in a standard language (e.g. with the help of computational models proposed by [28] or presented as a simulation task.
6. Running the algorithm.
   The input data should be chosen directly from the description of the problem pending analysis or deducted during its interpretation.
7. Evaluating effectiveness of the algorithm.
   At this stage, the algorithm outcome results are being interpreted in the context (terminology) of the task being solved.

8. Standardization of the developed algorithm.
    Preparing the solution algorithm token in the language of the repository (meta information in the form or an XML document).
9. Accumulating the algorithm token in the operational memory of the current training session.
10. Preparing the knowledge model in the form of an update to the repository.
    At this stage, the repository containing: a set of keywords – reflecting the content of the stated problem – from the domain knowledge models, task token and algorithm token, have to be filled.
11. Supplementing the existing repository.
    The required level of complement depends on the subject, goal and stage of training and has to be assigned by the teacher to every student.

8.4.3 The Method of Competence Verification in the Virtual Laboratory Environment

Analysis of figure 8.9 provides means to conclude two alternative scenarios. Firstly, if the student can adequately select a task - algorithm relevant to the presented problem in the context of virtual laboratory, the problem is addressed and subsequently solved. When the student is unable to immediately solve the presented problem, i.e. to appropriately identify the task - algorithm pair, he should be provided with tools allowing him to successfully that problem. In the following deliberations concerning competence verification, we will address the first situation since the objective at this stage is identifying domain-specific knowledge that the student is able to use and apply individually.

The competence verification executed in the virtual laboratory environment is premised on the idea of creating an interactive theoretical knowledge engineering environment. The task of creating such environment is of significant complexity because it requires a properly designed space for manipulating abstract knowledge structures and an outlet for visualizing theory laws and rules. The first activity upon creating the accurate environment is analysis of knowledge storage in conventional didactical materials. The analysis concludes that in order to relay theoretical knowledge a template is used. The template connects certain theoretical concepts with its pattern of use, both of which are usually presented using collection of examples, or use-cases. The student should be able to connect the portions of already acquired knowledge with ability to either solve the task, or apply a proper procedure, or algorithm. In addition, it should be mentioned that technical sciences in particular consider theoretical knowledge as a background for practical applications. Since the goal of the debated procedure is to analyze the already acquired individual's competencies, with respect to the presented material the remaining part of the chapter will present a method that is concerned with competence analysis of a student through investigating his ability to solve specific tasks. It can be concluded that it will strongly relate to his ability to connect theoretical knowledge with the requirements of the given problem.

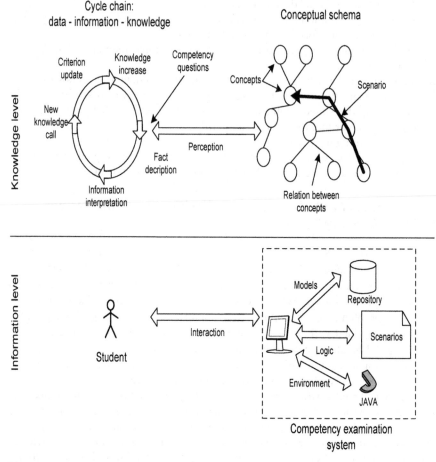

Fig. 8.10 The scheme of the competence verification procedures (source [24])

The presented method will be discussed in two-tier fashion (figure 8.10.). The knowledge level is defined by the knowledge range particular to human process of cognition discussed in works of [1]. Form cognitive science standpoint, the actions related to competence testing process can be interpreted as individual's recognition of the memorized patterns and articulate use of them. Through a cyclical operational data - information - knowledge the competencies are extracted form the individual's mind using competence question forms. The idea of competence question forms is in accordance with the approach used in ontology engineering formed [11]. The approach is characterized with an object containing domain specific knowledge that should be able to provide an accurate answer to a period. The competence question forms are mutually connected in a way that forms a scenario. Using such scenario we can see that a scenario is a mode of operation based on network-organized knowledge representation that is arranged to facilitate discovery of concept-to-concept relationships.

The level of information science presents a student that with a prepared interface manipulates the objects originating from the knowledge repository. That system is a topic-oriented database that is also capable of storing semantic concept structure and provide the knowledge portions in form of a model. The environment is implemented using popular programming language. Since the support of many computer platforms is required, use of Java technology seems effective.

An aspect of significant importance of the competence verification methods is a problem of forming the scenario of inter-related competence questions. Respecting the previously mentioned concerns dealing with the didactical material structure and the presented method of competence acquisition, the scenario has been based on the pool of competency questions (figure 8.11.). A structure of a competency question can be considered as an extensions to the knowledge repository based in the task - algorithm pair that is included within the competence acquisition procedure. The previously referred pair can be expanded to a tuple of conceptualization - problem - model - solution, where both the description and the problem are specialized forms, respectively, of the task and the algorithm.

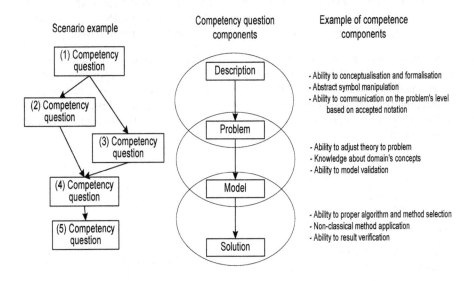

Fig. 8.11 The Relationships between the scenario components (source [25])

8.4.4 Implementation of the Method of Competence Verification

Using the proposed approach, a prototype computer program has been developed that satisfies the presented method of competence testing [25]. Upon the proceeding with the software development the following assumptions have been identified:

– ability to personalize the knowledge sharing environment;
– expandability of the remote knowledge base

- the users have predetermined roles and - depending on their properties - different permissions to access resources
- the program GUI is graphically appealing and ergonomic in design
- the system is able to handle and serve multiple concurrent students requests real time
- adaptive scenarios of knowledge relay.

The implemented program is based on client-sever architecture (please refer to figure 8.12. for more details). The client's application (non stand-alone) operates as an analogue of a terminal and operates on any computer. After successful login and authorization and depending on particular circumstances, all the missing data are retrieved from the server. We have used the following implementation guidelines:

- the software is platform-agnostic and therefore it operates using Java J2SE version 1.5.
- global access to the repositories is provided using web protocols such as SOAP, HTTP and technologies such as HTML, JSP, servlets and XML.
- particular elements have been formed using XML. Thanks to XSLT it is possible to easily convert the data to any desired and optimal presentation format.

The application server is comprised of database storing information about students, recording their results and overall progress and a repository storing determined problems, i.e. collections of data defined as ordered tuples (conceptualization, problems, models and solutions). At any desired time the repository can be supplied and expanded with new units formalized using the same formula.

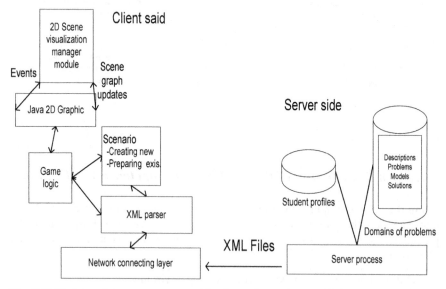

Fig. 8.12 Modules of the competence verification software (source [25])

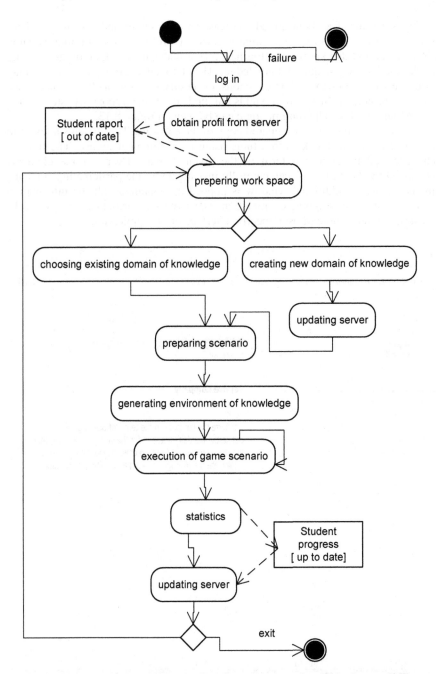

Fig. 8.13 An UML activity diagram, showing the software runtime logic (source [25])

The software of a client aimed to connect to the server and obtain a student-specific profile including accordingly selected problems for training session. Every scenario (also known as a tournament) provides feedback on progress being made and already acquired knowledge. The automated client application may insert new problems for training sessions, or modify the previously obtained ones. Every tournament finalization yields an update of the repository storing the individual's profile. The runtime logic can be seen on figure 8.13.

Figure 8.14. illustrates a prepared copy of the software. The students attempt to associate particular blocks into a tuple using their computer mouse. The tuple is characterized by a chosen term of the given domain. Every of the elements contained by the large area can carefully investigate by the participating students. In such case, an additional window is popped up containing all the information related to object undergoing examination. The screenshot presented on figure 8.14 illustrates and example of user-specific interface of the application.

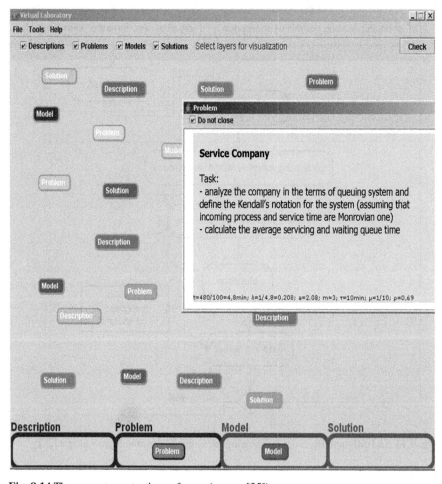

Fig. 8.14 The competence testing software (source [25])

8.5 Conclusions

One of the criterions for assessing the quality of the open distance learning process is the level of the competencies acquired by a student in a given time frame. The whole process of adapting both the tutoring programs and the didactical process to the changes occurring within the environment of the educational organization can be interpreted as an example of management oriented to maintain the sequence of information - knowledge - competence. In order to model the interaction of the experts involved with preparation of profiles and curricula, methods of game theory can be successfully applied.

The transfer of the didactical process into digitalized and networked environment and swapping of the direct interaction with a teacher for didactical materials requires accurate distinction of kinds of knowledge that is relayed to learning participants. Speaking in terms of competencies we have successfully distinguished: theoretical knowledge as a foundation for the object of study, procedural knowledge as technological underpinnings for completing practical tasks and, lastly, project-specific knowledge serving as a foundation for theoretical knowledge use for given task, or project in a given technological environment. The example illustrating that distinction between knowledge types is the virtual simulation laboratory.

The underlying element of the virtual laboratory is a repository that stores encoded theoretical knowledge related to particular stages of the simulation experiment (i.e. procedural knowledge). The repository enables performing recurrent automated transfer of fundamental knowledge to the students participating in the course of computer simulation. Using the virtual laboratory example, we have demonstrated a methodology for defining simulation models using the ARENA software that particularly designed for modeling queuing systems. Preparation and execution of the laboratory course using the proposed approach requires evolution of roles of all didactic process stakeholders that requires time cost and efforts. However, we expect that it is worth of trying as it clearly enriches and broadens the student's cycle of experiencing - being capable - succeeding.

References

1. Anderson, J.R.: Cognitive Psychology and Its Implications, 5th edn. Worth Publishing, New York (2000)
2. Banks, J., Carson, J.S., Nelson, B.L., Nivol, D.M.: Discrete-Event System Simulation. Prentice Hall, New York (2001)
3. Bourne, J., Harris, D., Mayadas, F.: Online Engineering Education: Learning Anywhere, Anytime. Journal of Engineering Education 94(1), 131–146 (2005)
4. Chung, C.A.: Simulation Modeling Handbook - A Practical Approach. CRC Press, Boca Raton (2004)
5. Conole, G., Dyke, M., Oliver, M., Seale, J.: Mapping pedagogy and tools for effective learning design. Computers & Education 43(1-2), 17–33 (2004)
6. Danna, K., Griffin, R.W.: Health and Well-Being in the Workplace: A Review and Synthesis of the Literature. Journal of Management 25(3), 357–384 (1999)

7. Dobrowolski, D.: Viral laboratory: new dimenssion of distacne learning. In: Straszaka, A., Owsiński, J. (eds.) BOS 2004, pp. 247–255. Akademicka Oficyna Wydawnicza EXIT, Warszawa (2004)

8. Duffy, V.G., Wu, F.F., Ng, P.P.W.: Development of an internet virtual layout system for improving workplace safety. Computers in Industry 50(2), 207–230 (2003)

9. García-Luque, E., Ortega, T., Forja, J.M., Gómez-Parra, A.: Using a laboratory simulator in the teaching and study of chemical processes in estuarine systems. Computers and Education 43(1-2), 81–90 (2004)

10. González-Castaño, F.J., Anido-Rifón, L., Valez-Alonso, J., Fernández-Iglesias, M.J., Llamas Nistal, M., Rodríguez-Hernández, P., Pousada-Carballo, J.M.: Internet access to real equipment at computer architecture laboratories using the Java/CORBA paradigm. Computers and Education 36(2), 151–170 (2001)

11. Gruninger M., Fox M.S.: Methodology for the Design and Evaluation of Ontologies. In: Proceedings of the Workshop on Basic Ontological Issues in Knowledge Sharing, IJCAI-1995, Montreal (1995)

12. Hwang, W.Y., Chang, C.B., Chen, G.J.: The relationship of learning traits, motivation and performance-learning response dynamics. Computers & Education 42(3), 267–287 (2004)

13. Knudsen, C., Naeve, A.: Presence Production in a Distributed Shared Virtual Environment for Exploring Mathematics. In: Sołdek, J., Pejaś, J. (eds.) Advanced Computer Systems: 8th International Conference, pp. 149–161. Kluwer Academic, Dordrecht (2002)

14. Kusztina, E., Zaikin, O., Różewski, P.: On the knowledge repository design and management In E-Learning. In: Lu, J., Da Ruan, Zhang, G. (eds.) E-Service Intelligence: Methodologies, Technologies and applications. SCI, vol. 37, pp. 497–517. Springer, Heidelberg (2007)

15. Kusztina, E., Zaikine, O., Różewski, P., Tadeusiewicz, R.: Competency framework in Open and Distance Learning. In: Proceedings of the 12th Conference of European University Information Systems EUNIS 2006, Tartu, Estonia, June 28-30, pp. 186–193 (2006)

16. Kusztina, E.: Conception of Open Information System for Distance Learning, Szczecin University of Technology. Faculty of Computer Science and Information Systems (2006)

17. Kushtina, E., Różewski, P., Zaikin, O.: Extended ontological model for distance learning purpose. In: Reimer, U., Karagiannis, D. (eds.) PAKM 2006. LNCS (LNAI), vol. 4333, pp. 155–165. Springer, Heidelberg (2006)

18. Kusztina, E., Dolgui, A., Małachowski, B.: Organization of the modeling and simulation of the discrete processes. In: Saeed, K., Pejaś, J. (eds.) Information Processing and Security Systems, pp. 443–452. Springer, Heidelberg (2005)

19. Law, A.M., Kelton, W.D.: Simulation Modeling and Analysis. McGraw-Hill, Boston (2000)

20. Popov, O., Barcz, A., Piela, P., Sobczak, T.: Practical realization of modeling an airplane for an intelligent tutoring system. In: Sołdek, J., Pejaś, J. (eds.) Advanced Computer Systems: 9th International Conference, pp. 149–161. Kluwer Academic Publishers, Dordrecht (2003)

21. Rak, R.J.: Virtual Instrument – the Main Part of Internet Based Distributed System. In: SSGRR 2000, L'Aquila, Italy (2000)

22. Ramasundaram, V., Grunwald, S., Mangeot, A., Comerford, N.B., Bliss, C.M.: Development of an environmental virtual field laboratory. Computers and Education 45(1), 21–34 (2005)

23. Robinson, S.: Simulation: The Practice of Model Development and Use. John Wiley & Sons, Chichester (2004)

24. Różewski, P., Kusztina, E.: Concept of competency examination system in virtual laboratory environment. In: Vossen, G., Long, D.D.E., Yu, J.X. (eds.) WISE 2009. LNCS, vol. 5802, pp. 489–496. Springer, Heidelberg (2009)
25. Różewski, P., Różewski, J.: Method of competency testing in virtual laboratory. In: Urbańczyk, E., Straszaka, A., Owsiński, J. (eds.) BOS 2006, pp. 349–360. Akademicka Oficyna Wydawnicza EXIT, Warszawa (2006) (in Pollish)
26. Scanlon, E., Colwell, C., Cooper, M., Di Paolo, T.: Remote experiments, re-versioning and re-thinking science learning. Computers and Education 43(1-2), 153–163 (2004)
27. Shackel, B.: People and computers - some recent highlights. Applied Ergonomics 31(6), 595–608 (2000)
28. Tyugu Enn, C.: Programming with knowledge base. WNT, Warszawa (1989) (in Polish)
29. Zaikin, O., Kusztina, E., Różewski, P.: Model and algorithm of the conceptual scheme formation for knowledge domain in distance learning. European Journal of Operational Research 175(3), 1379–1399 (2006)
30. Zaikin, O., Kusztina, E., Rozewski, P., Malachowski, B., Tadeusiewicz, R., Kusiak, J.: Polish experience in the didactical materials creation: the student involved in the learning/teaching process. In: Mahnic, V., Vilfan, B. (eds.) Proceedings of the 10th International Conference of EUNIS, IT Innovation in Changing World, , pp. 428–433. University of Ljubljana (2004)
31. Zaikin, O.: Queuing Modelling Of Supply Chai. In: Intelligent Production. Informa, Poland, Szczecin (2002)

Chapter 9
The Distance Learning Network

9.1 Introduction

The idea of an Intelligent Open Learning System assumes development of distance learning systems in the direction of open, personalized systems based on knowledge management tools. The aim of such systems is to support education on the level of passing knowledge, through creating of a computer environment dedicated to a specific student. Knowledge management methods allow for creation of knowledge portions adjusted to the education goals and the basic knowledge of the student. Simultaneously the openness of the environment enables creation of collaboration networks between students sharing similar education goals. One of the examples of such a system is the Distance Learning Network (DLN) [39].

Geographical dislocation of knowledge sources and students is a reason for development of an e-learning system into the DLN [39]. In such a network the nodes actively process knowledge and edges represent channels for knowledge relocation [5,10].

The major reason for creating a DLN is to provide students with access to different services created and based on internet. According to the results of the UE TENCompetence Project [24,26] a learning network connects actors, both human as well as agents, institutions and learning resources. Fulfilling such a function requires advanced computer and network infrastructure. Additionally, for a DLN to work effectively a good, corrected role distribution between social agent and proper access to services and tools are needed.

When creating a DLN we need to consider several aspects that arise from its complex character. The first important development aspect is the phenomenon of networked knowledge processing, the consequence of which are different roles of network users. Individual network nodes have different properties, what influences their social characteristics. Studying the social aspect of DL networks is one of the main conditions for building an effective environment of knowledge exchange.

The DLN can be analyzed from different points of view (Tab 9.1). Seeing the DLN as a telecommunication network, we concentrate mostly on the productivity of the network and the efficiency of data transmission. The point of view represented by workflow network defines efficiency of intangible production realized during the learning-teaching process. The following are the characteristics of intangible production realized in DLN [48]:

P. Różewski et al.: Intelligent Open Learning Systems, ISRL 22, pp. 213–238.
springerlink.com   © Springer-Verlag Berlin Heidelberg 2011

-   information character of the production process,
-   distributed character of production and product delivery,
-   production oriented on individual customer.

In our studies we focus on perceiving the DLN as a knowledge network. In such a network, knowledge is created and present through interactions between network participants. It is an important part of the DLN to study the social aspect of its functionality.

Table 9.1 Comparison of different analysis approaches to distance learning network (source [39])

Design concept	Telecommunication network	Workflow network	Knowledge network
Analytical basis	Queuing theory	Graph theory	Social network and ontology theory
Network's unit	Packet	Task	Knowledge/information object
Control parameters	Security, speed	Efficiency, workload	Completeness, credibility
Node concept	Computer station	Work station	Human-computer interaction station
Node role	Signal regeneration, data distribution	Technological operation	Knowledge and information distribution and creation
Work paradigm	Standards	Technological chain	Emergency, synergy

The concept of DLN presented in this chapter is different from other ones present in literature in its approach to treating DLN as a community-built system. The DLN is the operational environment of a community-built system and it allows processing of the networked knowledge. This approach assumes such selection of users, from the point of view of competences and education goals, that enables them to generate new content and support each other in their learning goals. It is also an objective of the DLN to enable users to easily share knowledge across the network. By sharing knowledge, an individual's expertise can be integrated to form a much larger and valuable network of knowledge [22], such as Wikipedia. The social-rooted approach to manage the Distance Learning Network can be used to optimize collaboration between the network's actors in order to efficiently management the knowledge flow. As a result of the analysis of the DLN from the social point of view, a motivation model was proposed. Afterwards, authors presented its application in the DLN working as a community-based system based on a knowledge repository. The main task of such a network is to create content. Our research was limited to the repository information system.

9.2 Knowledge Bases for the Distance Learning Network

Creating DLN requires concentration not only on the technological aspects. The central construction problems of DLN are placed in the area of knowledge management. We need to consider cognitive characteristics of the student and his/her social placement in the ecosystem of the entire network. Additionally, the described process has a cognitive character and is based on pedagogical principles. For the next analysis the DLN will be treated as a knowledge network.

A knowledge network gives information on "who knows what" [37], in other words it can be defined as a network in which all the knowledge of a given organization is properly represented, correlated, and accessed [32]. Operations of the knowledge network include knowledge sharing and knowledge transfer [9]. Knowledge sharing relates to the depth of the shared knowledge; knowledge transfer includes transport, absorption and feedback of knowledge.

The structure of the knowledge network in an organization (such as e.g. an education organization) is based on a production network and a social network that are directly related to the character of the workflow in the organization [9]. On the level of the social network, a knowledge node is a team member, a role, a knowledge portal or a process.

Relaying of knowledge between the nodes of the knowledge network is done according to certain production network rules and principles, and is referred to as knowledge flow [51]. Knowledge is, naturally, the main resource of every knowledge flow. Each knowledge flow starts and ends in a defined network node, which can generate, learn, process, understand, synthesize, and deliver knowledge [50]. The nodes in the knowledge network include individuals as well as aggregates of individuals, such as groups, departments, organizations, and agencies. Increasingly, the nodes also include non-human agents, such as community-built engines, knowledge repositories, web sites, content and referral databases [7].

9.2.1 Networked Knowledge Processing

Using [38] as our base, we can say that the processes of knowledge creation and of knowledge processing are both based on a person's individual cognitive capabilities, and on his/her access to resources (i.e. the object of processing) [38]. The aim of the networked organization of knowledge (networked knowledge) process is to ensure access to knowledge resources. It is also connected to a cognitive responsive environment, in which knowledge is created and processed. Networked knowledge can produce units of knowledge with informational value far beyond the mere sum of the individual units, i.e. it creates new knowledge [14]. The information resources available in the internet paired with appropriate semantics constitute the networked knowledge.

The phenomenons of synergy and of emergence are the most significant purposes for processing networked knowledge [38]. Synergy occurs when individuals with similar knowledge resources are mutually related. As a result of the relation, the initial knowledge potential and processed knowledge quality increase. Emergency occurs when individuals with similar objectives are connected. In this case, connecting resources of different roots in order to solve a task or a project allows for analyzing it from the points of view of different domains.

9.2.2 User's Role

Every node of the DLN is a social agent whose aim is to create new knowledge according to its logic. Each social agent has access to knowledge through:

– a personalized learning course,
– relations and contacts with other students,
– other active network resources containing an interface that allows for communication on the level of knowledge (e.g. knowledge repositories).

The reasons for new knowledge creation depend on the learning/teaching objectives. According to the concept of knowledge flow [39], social agents (social network vertex) plays the role of a knowledge portal or a knowledge process with a difference intensity. The node can generate, learn, process, understand, synthesize, and deliver knowledge [39]. Complete integration of different knowledge sources is possible basing on e-learning standards e.g. SCORM, IEEE LOM, CORDRA.

The nodes of the DLN can be interpreted as:

– knowledge workers (knowledge individuals) [30]
– knowledge systems [22].

The quality of the knowledge network is affected by connection schemas of the actors. Typical mistakes in maintaining connections (as described in [22]) include the following:

– actors connected only to actors with low expertise levels,
– actors receiving knowledge only by means of knowledge transfer with low viscosity,
– loosely connected sub-groups that do not profit from knowledge found in other groups,
– few actors preserving the entire network – without them only loosely connected subgroups would remain,
– actors not well integrated in the network – maintaining no or only a few relationships with other actors.

Each node (actor) in the DLN plays some role. As a response to network demands, the role an actor performs in the knowledge network may change. Each actor requires, desires and demands resources adequate to his intellectual potential, available, and in a proper network context. Based on [22], the following generic roles can be identified in the knowledge network:

– "knowledge creator" – an actor creating new knowledge that is used by others in the organization;

- "knowledge sharer" ("knowledge broker") – an actor responsible for sharing knowledge that is created by the knowledge creators;
- "knowledge user" – an actor who needs knowledge to perform a task.

The role of the knowledge broker in the community-built system is maintained by the system, which creates the operational environment for the knowledge network. In this system, additionally the role of an editor is introduced. The editor is responsible for estimating and maintaining quality of content in the knowledge network.

In the e-quality project [18], the following roles related to the learning-teaching process were recognised: Instructional designer, Content planner (Planning process); Educational Administrator, Coordinator, Technical administrator, Adviser (Administration process); Students' evaluator, Developer (Evaluation process); Material designer, Material producer, Audio-visual specialist (Learning material production process); Pedagogical support, Technological support, Tutor, Teacher, Student (Student's support process). Additionally, the e-learning system services also represent DLN's node e.g. repository, CMS, systems services (dictionaries, encyclopedias), mail services.

Relations between DLN nodes are the result of different aspects of the learning-teaching process in an educational organization. Basing on the eQuality report [19] almost 400 relationships can be distinguished (tab. 9.2. shows some examples).

Table 9.2 Examples of relationships in the DLN

Node 1	Content of relationships	Node 2
Instructional designer	To coordinate the production of the different learning objects, units and modules	Learning Management Systems
Instructional designer	To set up a plan and an implementation of a procedure of reusable components for further training actions	Best practices
Author	To provide references such as books, articles, websites…	Quotation software
Graphic designer	To create the illustrations	Author
Teacher/Counselor	To lead the learning dynamic	The didactic material
Tutor	To manage conflicts between students	Students
Head of the diploma	To choose the staff involved in each course, considering their attending capabilities	Learning plan
Technologic support	To guidance student's for the technology acquiring	Content Management Software

For every role recognized in the community-built system environment, a set of skills and behaviors is required [47]:

- cognitive skills: writing and constructive editing skills (skills in research, writing, and editing), web skills (accessing the Internet, using Web browsers, tracking logins and passwords, writing with embedded wiki HTML editors, and working with digital images or other Web media), group process skills (be able to set goals, communicate clearly, share leadership, participation, power, and influence, make effective decisions, engage in constructive controversy and negotiate conflict).
- personal characteristics: openness (open to each contributor's ideas, to scrutiny and criticism, allows others to modify, reorganize and improve any contributions), integrity (accountability, honesty, competence of each student's contribution), self-organization (requires the ability of metacognition, self-assessment, and the ability to adjust to feedback from the environment).

9.3 Community-Built System as a Main Paradigm for DLN

All of our discussions regarding the community-built system refer to, base on and use information from [38]. The community-built system can be defined as a system of virtual collaborations organized to provide an open resource development environment within a given community [41]. Virtual collaborations, according to [45], are a type of collaborations in which individuals are interdependent in their tasks, share responsibility for outcomes, and rely on information communication technology to produce an outcome, such as shared understanding, evaluation, strategy, recommendation, decision, action plan etc. A common goal or deliverable are needed in order to create the community-built system [36]. In this case, the common goal is to develop open resources according to the requirements of the learning-teaching process. An approach based on the community-built system to DLN allows using an education organization not only for knowledge distribution, but also for creating high-level content through the community's work.

The operational environment of a community-built system is defined as a collaborative service provided within a collaborative workplace, supporting collaborative activities of a collaborative group [23]. Collaborative systems, groupware, or multi-user applications allow groups of users to communicate and cooperate on common tasks [43]. Such systems are modelled as dynamic and inter-independent, diverse, complex, partially self-organizing, adaptive and fragile [6]. The community-based system operates using the asynchronous pattern, what is considered as promoting and supporting cooperation rather than competition among its users [16]. Basing on [11] e can formulate key assumptions necessary for a successful community-built system:

- knowledge is created and it is shared,
- individual's have prior knowledge they can contribute during a discussion/collaboration,

– participation is critical to open resources development,
– individuals will participate, if given optimal conditions.

9.3.1 The Collaboration Process

Over periods of time ranging from seconds to entire human generations many activities take place among participants in the field of learning, education and training. As was shown in [23], these activities are collaborative in nature, involve turn-taking, statement-and-response or multi-thread discussions. Usually people are driven to collaborate with others because they are given tasks that they cannot perform alone [36]. The community of collaborators is accelerated by social context, motivational aspects, distributed cognition, learning community [29]. One of the main purposes of the community-built system is to enable and promote collaboration. In this case, the social aspect of people gathering in a community becomes the most important one. The main collaborative knowledge processes here are construction and co-construction of knowledge, and reciprocal sense making [20].

The collaboration process is the driving force behind the community-built system's operation. It is important to distinguish between processes involving collaboration and coordinated work [2]. In collaboration no control element is required, all participants work together as partners (in opposite to a subordinate – supervisor scenario). Collaboration between individuals strongly depends on their relationship's typology [15]. Maintaining an all-to-all connection pattern is almost never possible. More likely, the connections pattern is limited to providing regular graphs, lattices, or other similar formats.

9.3.2 Wiki Community-Built System Engine

Community-built engine is, from the point of view of the community-built system, a hub in the production network. It is also the most important element of a social network representing the knowledge community. Wiki, a member of the asynchronous communication tools class, is an example of such an engine (Tab. 9.3.). It is a web-based hypertext system which supports community-oriented authoring, allowing authors to edit the page online in a web browser, thus enabling them to rapidly and collaboratively build the content [17].

Technically, wiki systems consist of four basic elements [25]:

– content,
– template which defines the layout of the wiki pages,
– engine, which handles all the business logic of the wiki,
– a wiki page – the page that is created by the wiki engine displaying the content in a web browser.

Wiki server technology enables creation of associative hypertexts with non-linear navigation structures [17].

Table 9.3 Main features of the Wiki and the community-built systems (source [38])

Wiki features (in the terms used by [44])	Wiki (based on [21])	Community-built system	Community-built system features
Open	If any page is found to be incomplete or poorly organized, any reader can edit it as he/she sees fit	Each member can create, modify and delete the content	Open source
Organic	The structure and content of the site evolve over time	The structure and content are dynamic and non-linear and can be rapidly constructed, accessed and modified	Rapidness
Universal	Any writer is automatically an editor and organizer	Every member can shift roles on the fly	Simplicity
Observable	Activity within the site can be watched and revised by any visitor to the site	System maintains a version database, which records its historical revision and content	Maintainable
Tolerant	Interpretable (even if undesirable) behavior is preferred to error messages	The content must fulfill interface requirements, the quality is decided by the community	Self-repair

9.4   Social-Based Approach to DLN Analysis

The organization of the learning-teaching process requires analysis of the student – teacher – e-learning system ecosystem, which can be based on different social-based methods, like Social Network Analysis. The social-rooted methods give possibility to interpret the DLN in the context of dynamics and topology of student's (agent's) communication. Moreover, a social model records the characteristics of the relations and interactions which occur between the DLN's elements (actors, services, etc.).

Generally speaking, the social-based approach allows to discover, basing on the shared social context and social artefact, authority relationships, information exchange, and affection or informal structures of the social network. Moreover, [46] gives several examples that the social-based approach is an efficient tool for network measurements and methodologies for evaluating virtual network communication properties, including central actors, short paths, core versus periphery, and clusters. Further on the social-based approach will be defined as an approach based on a social network. It is done in such a way in order to show the meaning of the network relations in the analysis of the social aspect.

9.4.1   Social Network Application in E-Learning

In e-learning the concept of social network is applied in two ways. The first approach is to implement the social network as a tool for e-learning system [12]. Such tools are used to extend Learning Management System about educational social software [3] i.e. networked tools that support and encourage individuals to learn together while retaining individual control over their time, space, presence, activity, identity and relationship.

The second approach of social network application in e-learning is using the Social Network Analysis (SNA) as an instrument for discovering interconnections, patterns or relationships between students and/or other learning process participants (tab. 9.4.). The SNA approach allows, among other things, for quantitative analysis of students' relationships [13].

Table 9.4 Review of social network application in education (source [39])

Idea	Focal point
Relations between social network characteristics in an online class and **cognitive learning outcomes**	Collaborative Learning environment, sense of community, prestige
Social network analysis provides meaningful and quantitative insights into the quality of the **knowledge construction process**	Knowledge process optimization, knowledge flow
Interactions in social groups and the impact of intrinsic characteristics of individuals on their **social interactions.**	Social architectures

What is more, e-learning institutions and specialists created several social networks used in educational environments or for educational purposes [42]. In current e-learning research the SNA tools and methods play only a supporting role. The results of SNA have not had direct impact on the operational level of the e-learning systems. There is a need for an approach to incorporate SNA tools and methods into the management process of the DLN.

9.4.2   Social Network Features in Distance Learning Network

Interpreting the DLN in terms of a social network allows us to assume that it is an undirected network, due to the intellectual relationships that occur between nodes. The direction of an intellectual relationship is not important as long as knowledge is transferred. The number of nodes N depends not only on the number of students, teachers, but also on the number of other active services and systems. Their activity is a result of their capabilities to process knowledge [50]. The number of edges is the result of the learning-teaching process organization.

For every node, the clustering coefficient [33] (including degree centrality, closeness centrality, betweenness centrality) can be calculated. The clustering coefficient metrics represent 'close' is neighbourhood of node i to being a clique (i.e. class of students) [33]. Such metrics allow estimating the importance of each node in the network. The node's role should be related to its clustering coefficient. High clustering coefficient for node i means the node i is an animator or a creator of knowledge in the DLN. High average clustering coefficient in the DLN assures high level of network socialization.

Degree distribution k_i of node i is the number of edges incident on (i.e., connected to) that node. Most of real life social networks are characterized by power-law degree distributions [1]. The same characteristic should be preserved in the DLN. The knowledge creator or animator should be the central point of every clique (class, common interest group). In order to fulfil such a postulate the DLN should be a scale-free network [4].

Analysis of the geodesic distance $l_{i,j}$ between nodes i and j gives an idea about their relationship's strength. Members of the current group of nodes (course students) should have the shortest distance between them. Moreover, the average geodesic distance $<l>$ between pairs of nodes in the DLN should be small as well. The reason for that is the high possibility that the connected students have similar knowledge. The DLN has to fulfill the Small World postulate [1], what means that almost every element of the network is somehow "close" to almost every other element, even those that are perceived as likely to be "far away". The Small World effect in the DLN ensures a fast access to all kinds of knowledge and resources and a rapid knowledge transfer.

9.5 Motivation Model as a Result of Social Characteristics of the DLN

9.5.1 Social Network Origin of the Motivation Model

The DLN as a social network is a social structure which connects individuals by one or more specific types of relationships related to the learning-teaching process. Basing on [17,25] we can differentiate the following types of relations:

- "who knows who" (classic Social Networks)
- "who thinks who knows who" (Socio-Cognitive Networks)
- "who knows what" (Knowledge Networks)
- "who thinks who knows what" (Cognitive Knowledge Networks)

The DLN combines in itself the characteristics of all the types of networks mentioned above. Analysis of the DLN from the point of view of a social network allows us to draw the following conclusions:

1) The most important resource of the DLN are its participants, especially students and teachers – they are the ones responsible for generating and distributing knowledge.
2) Each participant of the DLN can play different roles that influence his success in the network. Additionally, the importance of a resource (participant) in the network is also influenced by internal resources, e.g. possessing knowledge from a given domain at a high level.
3) Effectiveness of DLN functionality depends on different types of relations that occur between its nodes.

In this context, the motivation model is the approach to managing resources of the DLN (e.g. new content) through influencing the internal motivation of the DLN's participants. Every DLN participant can be characterized by his/her motivation. The collective motivation model of users allows increasing the overall DLN productivity through managing at the level of relations and cooperation. It is a step forward in regard to the queuing model, which concentrates only on the productivity of the network as such. Motivation allows for influencing the engagement of DLN participants, what is reflected in their will to cooperate in the frames of the network. The motivation model is especially important in network structures where productivity depends on cooperation.

9.5.2 Motivation Model Concept

Developing a motivation model for different organizations and environments has long been a scientific research problem [27]. The problem of motivation in control in social economic systems is interpreted as an incentive problem [35]. The incentive problem relies on inducing the elements of organization to undertake certain actions. Literature shows that the incentive problem is explored in psychology, economy, sociology and management theory. As was shown in [34], the complexity of incentive problems is stipulated by the activity of the managed objects – agents, i.e. students. Each student is characterized by independent choice of states and actions, information misrepresentation, ability for rash behavior etc.

Stimulation, motivation (incentives) is a complex and purposeful influence on the components of agent's activity. Moreover the environment also influences on the agent. Motivation (stimulation) is usually considered as the influence on agent's tasks and processes of their forming with fixed needs, motives, purposes and technologies.

When building a model of a DLN participant we can assume that his actions are steered by needs. According to [35], the process of internal motivation which is based on the need leads to forming the motive and activity's purpose (anticipated result of activity). In case of the DLN, the need and the motive are defined by learning objectives and the social environment in the network. Following Novikov, the purpose being coordinated with external and internal conditions is transferred onto a set of tasks. Tasks are solved through application of certain technologies provided by the DLN: a set of content, forms, methods and means. The action connected to performing and realizing a task with the use of available technologies leads to some results (satisfaction or partial satisfaction of the need).

Comparison of these results with the purpose can lead to modification of the components of the performed action (tasks, technology, etc.).

9.5.3   Interpretation of the Motivation Model in the Context of the DLN

The motivation model supports cooperation at the level of creating content. One of the concepts of creating the DLN is activity based on a community-built system. The DLN can be treated as a social network that connects different communities. Like every community-built system, the DLN has a specified common goal. In case of the DLN, the community-built system goal is to develop open resources according to the requirements of the learning-teaching process.

The proposed motivation model is designed to support activity of editors and creators performed during the learning-teaching process. In the analysed situation, the DLN is treated as an information system working in concordance with the concept of community-built systems. In such systems resources are developed mainly through the work of the system's participants. Wikipedia is a good example of such a system.

In DLN, having in mind the process of creating content, we can identify two roles: creator and editor. Repository is the system that integrates them (figure 9.1.). The process of content creation always includes the social, research and education aspects and occurs across cognitive, information and computer-based levels of community-built systems. At each of these levels, the editors and the creators have their own roles and levels of involvement. In the community-built system, motivation plays a different part in the content creation and modification processes. The creator is a person responsible for creating and improving the content. The editor is a person responsible for content assessment.

The active actors take part in the system's working routine. The active (on-line) actors play the roles of creators (content creation) and/or editors (content evaluation and consultation) [27,38]. The passive actors are outside of the system, but they are registered in the system and can become active without any cost. During the passive (off-line) mode the creator is thinking and increases the knowledge potential and editor applied knowledge during the learning-teaching process.

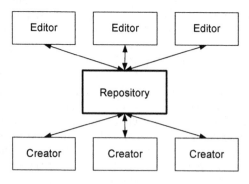

Fig. 9.1 Model of a community-built system

Table 9.5 Comparison of community-built systems with and without authority

	Community-built system with authority	Community-built system without authority
Role	Editor: teacher Creator: student	Editor: user Creator: author
Editor's interest	The editor's interest is in maximizing the level of repository filling with tasks of different complexity for every considered educational situation in a given domain. The criterion regarding placing a task in the repository is decided by the teacher on the basis of: complexity level, graphical quality, language correctness etc. The possibility to realize teacher's interests is limited by his/her resources considering time in the quantitative and calendar aspects and other informal preferences.	The editor's main motivation is such content verification, that it reflects his own understanding of knowledge in the content repository. Hence, the content is the editor's main interest focus. Therefore, the editor is responsible for development of the repository. The repository development is carried out by the editor's actions, such as selection of a concept for future improvement conducted by a given creator.
Editor's motivation function	The editor's motivation function is oriented on maximizing the span of the domain with tasks (concepts) that have the following qualities: topicality of the task's subject from the editor's point of view and the individual resources he is prepared to assign to the creator for solving a certain task (for example consultation time, access to scientific material, etc.). Maximizing the degree of repository filling with concepts is the editor's interest. It the end, the repository should reflect the complete ontology of a given subject.	
Creator's interests	The creator's interests relies on individual preferences and can be described using opposing groups of creators. The first group is concerned mainly with achieving a minimal acceptable success level, meaning meeting only the basic requirements for obtaining a positive opinion about the task (low complexity of the task, minimal acceptable quality) and saving maximal amount of their time. The second group of creators is interested in providing the community-built system repository with the maximal possible success level, implying creating and editing contents of high complexity in order to produce best overall quality.	
Creator's motivation function	The creator's motivation function considers obtaining maximum level of fulfilling one's interests during choosing and solving the task, with given constraints regarding time (one's own and the teacher's) and the way of grading the resulting solutions.	

Analysis of the repository-based content creation process that is part of the DLN allows for identifying two situations (Tab. 9.5.). The first one assumes that it is impossible to change the role in the network (figure 9.2). Individual nodes of the DLN play either the role of the editors or of the creators. Such an approach was called community-built system with authority. The second approach assumes that a node can, in a certain moment, be either the editor or the creator, and the roles can be changed (figure 9.3).

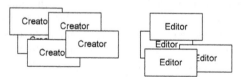

Fig. 9.2 Community-built system without changing roles

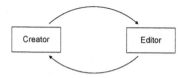

Fig. 9.3 Community-built system with changing roles

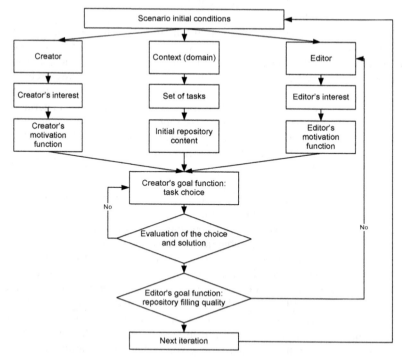

Fig. 9.4 The scenarios of supplying and using the community-built system repository with authority (source [27])

In the scenario for community-built system repository with authority, presented in figure 9.4, it is assumed that the role of the student is to choose a task and solve it. Depending on the correctness of the solution and the complexity level of the task, the final grade is established. A task highly graded by the teacher is placed in the repository and will serve as an example solution for other students.

Figure 9.5. represents the proposed scenario for a community-built system repository without authority, in which it is assumed that the role of the creator is to choose a task after the pre-processing phase. The purpose of the pre-processing phase is to estimate what the content of the repository should be from the point of view of the creator's interest (creator's preference). Every task is related to some part of the content development process, thus a new task represents a new concept.

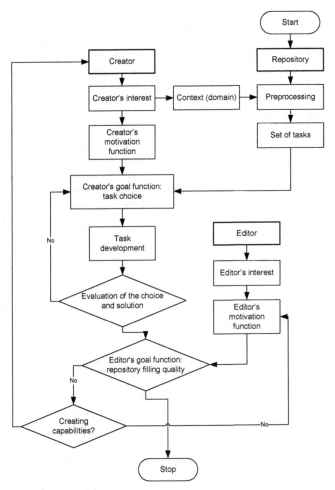

Fig. 9.5 The scenario of supplying and using the community-built system repository without authority (source [38])

The creator deals with the task, and the task quality depends on the correctness of the development process and the complexity level of the task. Both the task solved by the creator and the assessment accepted by the editor are placed in the repository and will serve as a base solution for other creators.

Formal approach to motivation model can be found in [27,38].

9.6 Example Approach to DLN Development

9.6.1 Model of a Collaboration Environment for Knowledge Management in Competence Based Learning

In this example of DLN both groups of actors - students and teachers - work jointly, to provide the knowledge repository with high quality content. This example is based on [40]. In that article, the authors propose a model of educational and social collaboration between the students, the teacher, and the e-learning information system (the repository), as an example of the social learning process [8]. The system supports the competence-based learning process by means of creating a social network in order to exchange the ontology in the repository environment. The student is included in the repository development process. The idea is to expand the student's knowledge during the learning process, and record his achievements using an external e-portfolio mechanism. The repository is used as a mechanism in the learning process, while the e-portfolio is an individual property of each student and represents his/her achievements in the learning process. In order to make competence-based learning process more effective it is necessary to introduce a new organization of the learning process and collaboration "student-information system-teacher". Figure 9.6. presents the collaboration model.

The proposed model of the collaboration environment is divided into three levels [40]:

1) knowledge level – deals with proper preparation of the learning process from the point of view of the structure of domain knowledge; the process of structuring knowledge includes preparation of portions of theoretical and procedural knowledge, and of corresponding test tasks, to verify students' skills/abilities, in accordance with the accepted definition of knowledge;

2) social network level – is the basis for determining conditions for cooperation between the teacher and the students; appropriate motivation functions described this cooperation, which creates a new dimension for social networks;

3) management level – is responsible for managing the cooperation; conditions of the cooperation are analyzed from the point of view of existing constrains (human and time resources), and the obtained results can signal a need to change the cooperation model. Feedback information is also used for possible further repository development and designing collaboration of the learning process participants.

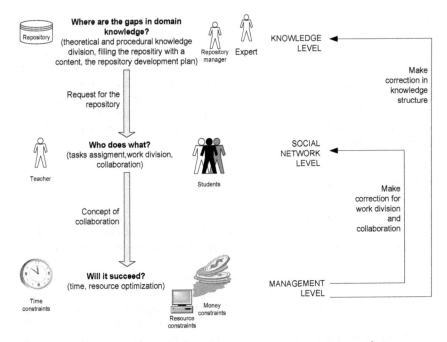

Fig. 9.6 Model of collaboration environment for knowledge management in competence-based learning (source [40])

Combination of these levels allows organizing the learning process in accordance with the requirements for competence-based learning and involving students, who become active and equal participants of the learning process and consciously create their own cognitive process.

9.6.2 Knowledge Level

The knowledge level is based on the heuristic algorithm presented in [28]. The main part of this algorithm concerns presenting domain knowledge in a way appropriate for a specific educational situation. The algorithm also allows dividing knowledge in accordance with the LOs philosophy, in other words by using knowledge modules in digital form, shared in a network, and reusable in a learning process [31]. More details about the knowledge division process and preparation of LOs sequence can be found in [49].

The methodology used for this purpose allow determining proportion and depth of domain knowledge in the specified course/subject. It is also necessary to assess the completeness of coverage of domain knowledge by the didactic materials. The starting point for the model is domain ontology and its division into theoretical and procedural knowledge.

The person responsible for the content the repository has to make sure that the complexity level of the elaborated topics is well differentiated and that the disproportion between the didactic materials content for the different topics is as small as possible.

The idea is to base the repository development on students' tasks. The task solution prepared by a student can be loaded into the repository when a different method for solving the given problem is used, or when there is a new task solved using a known method or the task and the solution represent a new research area but are in accordance with the goals and scope of the learning process. From the student's point of view the students solutions loaded into the repository can be and element of the e-portfolio and personal learning path development.

9.6.3 Social Network Level

The crucial element in forming cooperation in the learning process is basing the repository development plan on ontology. The social network created for the collaboration process consists of a teacher, students and the repository – an activity "accelerated" mechanism in this network.

Behaviour of the participants of the learning process can be described by appropriate functions of motivation. The teacher's motivation function aims at influencing the students in a way that will lead to filling the repository with new content. Their solutions are proof of the high competence level they acquired. The teacher's motivation function depends on the complexity of tasks, tasks relevance and other teacher's preferences. It can be described by a vector $\sigma^N(r_i) = \bar{x}(r_i)$, where $\bar{x}(r_i)$ - resources assigned to solving task r_i (like didactic materials, teacher's time, software, hardware, etc).

The student's motivation function is the result of the accepted development plan. Students decide about their own involvement in the learning process for a specified subject. This strategy depends on many factors:

- the repository content,
- tasks' complexity level,
- time needed to solve a task,
- potential grade/mark received for the tasks sequent execution.

For an set of tasks in the repository, the student's motivation function is developed individually for each student (S) and can be described as $\sigma_j^S(r_i) = F(W(s_j), H(r_i), C_j^S(r_i), F^S)$, where $W(s_j)$- student's base knowledge, $H(r_i)$- grade/mark, that can be given by a teacher $C_j^S(r_i)$- costs bear by the student regarding solving the task, F^S - other student's preferences (e.g. student's goals and constrains in the learning process).

The synergy effect and, on its basis, the repository enrichment with new content, requires finding balance between the teacher's motivation σ^N and the student's motivation σ_j^S.

The teacher's interests are satisfied when a properly solved task of a significantly high level of complexity is added to the repository. The student's interests are satisfied by minimal summary time costs of obtaining a high grade, which also depends on the complexity level of the task.

Balance between the teacher's and student's motivation functions defines the goal function of the task choice: $\Phi(y_j^i) = \alpha \sigma^N + \sigma_j^S = \max_Y, \quad i = 1,2,.....,i*, \quad j = 1,2,....j*$, where α is a weigh coefficient.

The motivation model can be formulated in terms of the games theory, as was shown in [27]. However, due to the fact that cooperation between the teacher and the student is conducted with time, financial and resource constraints, simulation tools (e.g. simulation tool Arena) can be used for its analysis.

9.6.4 Management Level

Ultimately, the repository should contain a set of didactic material units which completely cover the prepared ontology. Whether the content prepared by a student can be used to develop the repository will be determined by the teacher's assessment – only highly graded content will be used.

The effectiveness of this type of work with the repository depends on its initial state, the quantity of students involved in the development, the state of the desired repository level, meeting the time and resource utilization constraints. The accepted knowledge management and the learning process participants' motivation strategy requires a verification /calculation mechanism.

The analysis of the assumptions of the motivation model, their adjustment and the effectiveness of the repository development process, can be verified through the use of a simulation model. The simulation model allows for analyzing the parameters of the formalized learning process and the proposed cooperation between the learning process' participants. Assumptions such as time between students tasks arrival, time of the tasks assessment process or delay time, became the basis for the simulation model in Arena (figure 9.7.). Arena is a simulation tool developed by Rockwell Software, and allows for modeling and analyzing real life processes. Some aspects of using Arena for learning process analysis can be found in [40]. Currently the simulation experiment allows analyzing the queue's parameters on a teacher's workstation as well as defining the students' service time through specified input parameters. Figure 9.7. presents the results of such simulation.

For 55 students, time interval of 6 days, daily time for tasks' examination – 3 hours, expected time for each student – 20 minutes, correction time – 1 day and predicted probability of tasks' evaluation: 70% – exit with promotion without repository development, 15% – placing solution in the repository, 15% – sending back for correction; one can see that a queue on the teacher's workstation occurs. This queue shows that accepted assumptions for the learning process realization should be changed. As it was checked in the model, the time interval can be extended to 8 days to realize the other assumptions.

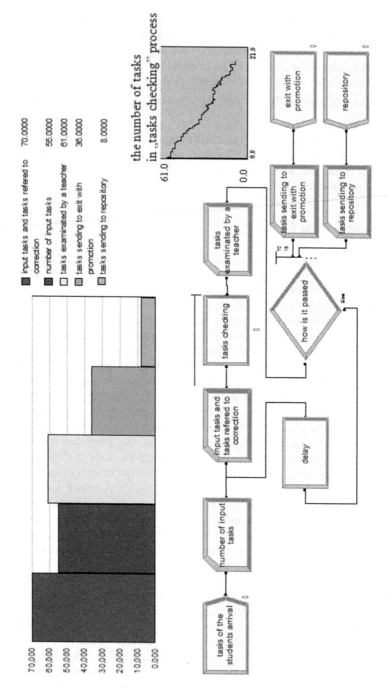

Fig. 9.7 Simulation model of teacher-students collaboration in tasks assessment (Arena software) (source [38])

Simulation also allows assessing: limited access to software and hardware resources, maximum size of social networks queue, cost of teacher's work. On the basis of statistical data, the group management strategy can be modified and appropriate adjustments can be made in the repository development plan.

9.7 The DLN as an Example of Intelligent Open Learning Systems

In order to fully support the concept of DLN in the Intelligent Open Learning Systems, the initial concept of OSDL has to be broadened. In previous chapters authors performed a soft system analysis based on the theory of hierarchical multilevel systems, which allows developing a model of an Open System of Distance Learning (OSDL). As a result of the soft system analysis, a hierarchical structure of the OSDL is proposed. The OSDL's hierarchical model incorporated the learning object approach to knowledge repository development, the social aspect of the e-learning process, and competence-based learning. The hierarchical structure of the information system defines the set of its sub-systems and their functioning scope. The following sub-systems are recognized: Learning Management System, Learning Content Management System, and Strategic Management System. For meeting the requirements of DLN a new module was added: Learning Social Management System (figure 9.8.).

Nowadays e-learning systems take the form of a distance learning network due to wide application of internet-based and networked e-learning services. In such networks the collaboration process become the most important one. The Learning Social Management Systems (LSMS on figure 9.9) class support the collaboration process through the social network approach. The LSMS supports and encourages individuals to learn together while retaining individual control over their time, space, presence, activity, identity and relationship. LSMS consists of three components:

- Social network support – system supporting the activity of DLN on the level of social network. It consists of systems responsible for functioning on the level of knowledge network (network of semantic relations between students) and production network (workflow defining the network structure of the learning-teaching process).
- Collaborative tools – a set of tools allowing communication and cooperation through supporting social collaboration. A typical example could be a Wiki or tools for group work.
- Competence profile support – systems and methods responsible for developing a social profile of a student in DLN. In the context of the classically understood social network, the accent should be placed mostly on the character or cognitive features. In the DL network, where the goal is learning and knowledge development on the basis of the community-built system concept, accent is placed on supporting the processes of creating and developing knowledge. The competence profile defines the scope of our capabilities and knowledge in the context of the given task or problem.

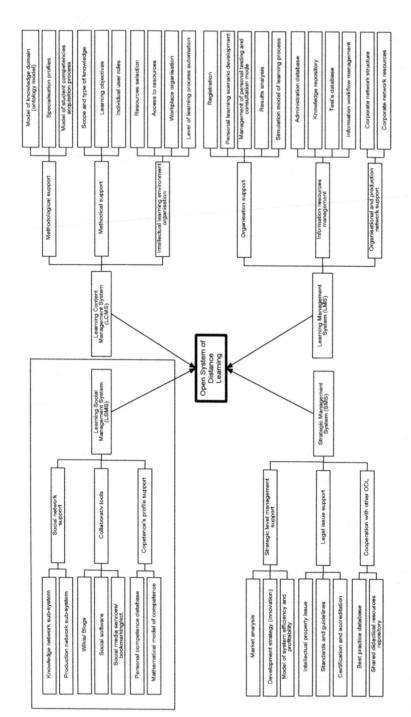

Fig. 9.8 Hierarchical structure of the Open System of Distance Learning

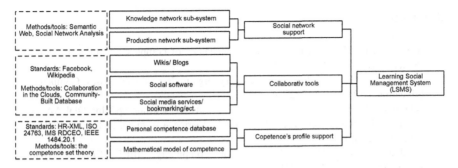

Fig. 9.9 Detailed view on the Learning Social Management Systems

9.8 Conclusions

In the chapter, the concept of DLN was presented. It combines the social character of the teaching-learning process with computer-based methods of knowledge management. In the proposed approach, DLN is the operational environment of the community-built system and it allows processing of networked knowledge. The approach proposes such a choice of users, from the point of view of competences and learning goals, that they can generate new content and support each other in realizing their learning goals.

We should take advantage of the fact that the modern distance learning system is networked. Apart form the research issues of distributed resources and work organization, one should put some attention to network characteristic of the learning process. The activity of the DLN can be analyzed both on technical and knowledge levels based of the motivation model. The motivation model covers two functions important for motivation: that of the creator and that of the editor, and describes their unique interests in supplying a knowledge repository.

In this chapter, the DLN has been identified as a form of a social network. In DLN the knowledge is modeled using the ontological approach, which allows to represent knowledge in a formal way available for further computer processing. Moreover, the ontological approach is compatible with the concept of Semantic Web. In the future, the DLN can be integrated with the Semantic Web. Subsequently, multiple software agents can become participants in the process of populating the knowledge repository.

The model of social agent collaboration between students, teacher and the e-learning information system (repository) allows for:

– assessing the degree of the repository fulfillment and its further development plan;
– checking the acquired competence path through appropriate domain knowledge structurization;
– measuring competence growth through analysis of the domain ontology graph coverage;
– elaborating a model of teacher behavior with successive groups of students;
– modeling and reviewing the group management strategy in accordance with teacher's and students' preferences.

References

1. Albert, R., Barabási, A.-L.: Statistical mechanics of complex networks. Reviews of Modern Physics 74, 47–97 (2002)
2. Alexander, P.M.: Virtual teamwork in very large undergraduate classes. Computers & Education 47(2), 127–147 (2006)
3. Anderson, T.: Distance learning – Social software's killer ap? In: ODLAA 2005 Breaking the boundaries: The international experience in open, distance and flexible education (2005)
4. Barabási, A.-L.: The architecture of complexity. IEEE Control Systems Magazine 27(4), 33–42 (2007)
5. Ben, C., Nien-Heng, C., Yi-Chan, D.,Tak-Wai, C.: Environmental design for a structured network learning society. Computer & Education 48(2), 234–249 (2007)
6. Brown, J.S.: Leveraging technology for learning in the cyber age – opportunities and pitfalls. In: International conference on computers in education, ICCE 1998 (1998) (invited Speaker)
7. Carley, K.: Smart agents and organizations of the future. In: Lievrouw, L., Livingstone, S. (eds.) Handbook of new media, pp. 206–220. Sage, Thousand Oaks (2002)
8. Chang, B., Cheng, N.-H., Deng, Y.C., Tak-Wai, C.T.W.: Environmental design for a structured network learning society. Computers & Education 28(2), 234–249 (2007)
9. Chen, J., Chen, D., Li, Z.: The Analysis of Knowledge Network Efficiency in Industrial Clusters. In: Proceedings of the 2008 International Seminar on Future Information Technology and Management Engineering (FITME 2008), pp. 257–260. IEEE Computer Society, Leicestershire (2008)
10. Cho, H., Gay, G., Davidson, B., Ingraffea, A.: Social networks, communication styles, and learning performance in a CSCL community. Computer & Education 49(2), 309–329 (2007)
11. Cole, M.: Using Wiki technology to support student engagement: Lessons from the trenches. Computers & Education 52(1), 141–146 (2009)
12. Dalsgaard, C.: Social software: E-learning beyond learning management systems. European Journal of Open, Distance and E-Learning 2 (2006)
13. Dawson, S., McWilliam, E., Tan, J.P.L.: Teaching smarter: How mining ICT data can inform and improve learning and teaching practice. In: Hello! Where are you in the landscape of educational technology? Proceedings of ascilite Melbourne (2008)
14. Decker, S., Hauswirth, M.: Enabling Networked Knowledge. In: Klusch, M., Pěchouček, M., Polleres, A. (eds.) CIA 2008. LNCS (LNAI), vol. 5180, pp. 1–15. Springer, Heidelberg (2008)
15. Delgado, J.: Emergence of social conventions in complex networks. Artificial Intelligence 141(1-2), 171–185 (2002)
16. De Pedro, X., Rieradevall, M., Lopez, P., Sant, D., Pinol, J., Nunez, L.: Writing documents collaboratively in Higher Education using traditional vs. Wiki methodology (I): qualitative results from a 2-year project study. In: International Congress of University Teaching and Innovation, Barcelona, July 5-7 (2006)
17. Ebersbach, A., Glaser, M., Heigl, R.: Wiki: Web Collaboration, 2nd edn. Springer, Heidelberg (2008)

18. E-Quality: Quality implementation in open and distance learning in a multicultural European environment. Socrates/Minerva European Union Project 2003-2006 (2006), http://www.e-quality-eu.org

19. E-Quality, A.: Conceptual Model for ODL Quality Processes. Deliverable 2.2 from Project, Downloadable form (2006), http://www.e-quality-eu.org/deliverable_2p2.html

20. Fischer, F., Bruhn, J., Grasel, C., Mandl, H.: Fostering collaborative knowledge construction with visualization tools. Learning and Instruction 12(2), 213–232 (2002)

21. Frumkin, J.: The wiki and the digital library. OCLC Systems & Services 21(1), 18–22 (2005)

22. Helms, R., Buijsrogge, K.: Knowledge Network Analysis: A Technique to Analyze Knowledge Management Bottlenecks in Organizations. In: Andersen, K.V., Debenham, J., Wagner, R. (eds.) DEXA 2005. LNCS, vol. 3588, pp. 410–414. Springer, Heidelberg (2005)

23. ISO 19778 Information technology - Learning, education and training - Collaborative technology - Collaborative workplace. International Organization for Standardization (2008)

24. Kalz, M., Van Bruggen, J., Rusmann, E., Giesbers, B., Koper, R.: Positioning of Learners in Learning Networks with Content-Analysis, Metadata and Ontologies. Interactive Learning Environments 15, 191–200 (2007)

25. Klobas, J.: Wikis as Tools for Collaboration. In: Kock, N. (ed.) Encyclopedia of E-Collaboration. IGI Publishing, Hershey (2008)

26. Koper, R., Rusman, E., Sloep, E.: Learning Network connecting people, organisations, software agents and learning resources to establish the emergence of effective lifelong learning. LLine: Lifelong Learning in Europe 9(1), 18–27 (2005)

27. Kusztina, E., Zaikin, O., Tadeusiewicz, R.: The research behavior/attitude support model in open learning systems. Bulletin of the Polish Academy of Sciences: Technical Sciences 58(4), 705–711 (2010)

28. Kusztina, E., Zaikin, O., Ciszczyk, M., Tadeusiewicz, R.: Quality factors for knowledge repository: based on e-Quality project. In: Joergensen D.S. & Stenalt M.H. (eds.): Proceedings of EUNIS 2008 VISION IT - Vision for IT in higher education, University of Aarhus (2008)

29. Lowyck, J., Poysa, J.: Design of collaborative learning environments. Computers in Human Behavior 17(5-6), 507–516 (2001)

30. Marwick, A.D.: Knowledge management technology. IBM System Journal 40(4), 814–830 (2001)

31. McGreal, R.: A Typology of Learning Object Repositories. In: Adelsberger, H.H., Kinshuk, P.J.M., Sampson, D. (eds.) Handbook on Information Technologies for Education and Training, 2nd edn., pp. 5–28. Springer, Heidelberg (2008)

32. Mulvenna, M.D., Zambonelli, F., Curran, K., Nugent, C.D.: Knowledge networks. In: Stavrakakis, I., Smirnov, M. (eds.) WAC 2005. LNCS, vol. 3854, pp. 99–114. Springer, Heidelberg (2006)

33. Newman, M.E.J.: The Structure and Function of Complex Networks, SIAM Rev. 45(2), 167–256 (2003)

34. Novikov, D.A., Shokhina, T.E.: Incentive Mechanisms in Dynamic Active Systems. Automation and Remote Control 64(12), 1912–1921 (2003)

35. Novikov, D.A.: Incentives in organizations: theory and practice. In: Proceedings of 14th International Conference on Systems Science, Wroclaw, vol. 2, pp. 19–29 (2001)

36. Ommeren, E., Duivestein, S.: deVadoss J., Reijnen C., Gunvaldson E.: Collaboration in the Cloud - How Cross-Boundary Collaboration Is Transforming Business. Microsoft and Sogeti (2009)

37. Palazzolo, E.T., Ghate, A., Dandi, R., Mahalingam, A., Contractor, N., Levitt, R.: Modelling 21st century project teams: docking workflow and knowledge network computational models. In: CASOS 2002 Computational and Mathematical Organization Theory Conference, Pittsburgh, USA, June 21-23 (2002)

38. Różewski, P., Kushtina, E.: Motivation Model in Community-Built System. In: Pardede, E. (ed.) Community-Built Database: Research and Development. Springer ISKM Series (2011)

39. Różewski, P.: A Method of Social Collaboration and Knowledge Sharing Acceleration for e-learning System: the Distance Learning Network Scenario. In: Bi, Y., Williams, M.-A. (eds.) KSEM 2010. LNCS, vol. 6291, pp. 148–159. Springer, Heidelberg (2010)

40. Różewski, P., Ciszczyk, M.: Model of a collaboration environment for knowledge management in competence-based learning. In: Nguyen, N.T., Kowalczyk, R., Chen, S.-M. (eds.) ICCCI 2009. LNCS, vol. 5796, pp. 333–344. Springer, Heidelberg (2009)

41. Shen, D., Nuankhieo, P., Huang, X., Amelung, C., Laffey, J.: Using Social Network Analysis to Understand Sense of Community in an Online Learning Environment. Journal of Educational Computing Research 39(1), 17–36 (2008)

42. Social Networks in Education, http://socialnetworksined.wikispaces.com/

43. Sugumaran, V., Storey, V.C.: Ontologies for conceptual modeling: their creation, use, and management. Data & Knowledge Engineering 42(3), 251–271 (2002)

44. Wagner, C.: Wiki: A technology for conversational knowledge management and group collaboration. Communications of the AIS 13, 265–289 (2004)

45. Wainfan, L., Davis, P.K.: Challenges in Virtual Collaboration: Videoconferencing Audioconferencing and Computer-Mediated Communications. RAND Corporation, Santa Monica (2005)

46. Wasserman, S., Faust, K.: Social Network Analysis: Methods and Applications. Cambridge University Press, Cambridge (1994)

47. West, J.A., West, M.L.: Using Wikis for Online Collaboration: The Power of the Read-Write Web. Wiley, Chichester (2008)

48. Zaikin, O.: Queuing Modelling Of Supply Chai. In: Intelligent Production. Informa, Poland, Szczecin (2002)

49. Zaikin, O., Kushtina, E., Różewski, P.: Model and algorithm of the conceptual scheme formation for knowledge domain in distance learning. European Journal of Operational Research 175(3), 1379–1399 (2006)

50. Zhuge, H.: Discovery of Knowledge Flow in Science. Communications of the ACM 49(5), 101–107 (2006)

51. Zhuge, H.: Knowledge flow network planning and simulation. Decision Support Systems 42(2), 571–592 (2006)

Chapter 10
AGH Student City as an Example of Open and Distance Learning System

10.1 Introduction

The idea of Open and Distance Learning System is considered as a highly interesting subject of scientific research and practical developments. However, if one chooses to confront the outcomes of theoretical considerations with reality, or attempts to quantify the effectiveness of the effectuated deployments, it is likely he or she will meet significant arising complications. The performance of Open and Distance Learning is above all a difficult subject for any experimental verification procedure, due to the foundational principle of such system almost entirely precludes any research proceedings. That is mainly a consequence of making educational resources publicly available which in turn renders impossible tracing their state, or gathering information on how and to what effect they are used.

On the other hand, it is rather difficult to transfer the outcomes of scientific observations and experiments to the Open and Distance Learning Systems environment as they are often published by individuals specialized in e-learning methods but using them in typical form and limited to a single student group, or - less likely - to a single term. The reasons for that limited outcomes transferability are usually the following:

- in e-learning courses, students are usually subject to a form of sanctioned regime since the learning process needs to be finalized with graduation or examination. In typical e-learning scenarios, that induces an obligation to use predefined digitalized educational resources. That very fact is contradictory to the idea of unrestrained accessibility of educational resources which itself is a foundation for the Open and Distance Learning process.
- apart from the obligation to learning, typical e-learning courses require framing the learning process within defined time frame that is subject to protocols of conducting, grading and finalization of the courses. That does not correspond with the Open and Distance Learning process that promotes learning in any given time and place.
- during teaching with use of e-learning as an aid to the primary process, a group of students (at school or university) is usually entirely homogenous, i.e., in the same age and with the roughly the same amount of previously acquired knowledge. However, the makeup of learning individuals participating in the Open and Distance Learning process is characterized by varying age and unequal initial knowledge.

P. Różewski et al.: Intelligent Open Learning Systems, ISRL 22, pp. 239–257.
springerlink.com
© Springer-Verlag Berlin Heidelberg 2011

The aforementioned reasons justify the assertion that accessibility of information required for evaluation and optimization of Open and Distance Learning Systems models is substantially limited. That holds true even when we consider the situation of increasing volume of research concerning methods and outcomes of applications of tele- and IT-assisted learning with respect to various e-learning forms. Meanwhile, the mounting determination of institutions and organizations involved in advocacy and implementation of Open and Distance Learning class systems necessitates an immediate and urgent requirement for both the evaluation procedures and the premises for optimization.

Having strong considerations for the problems so far presented herein we should relate with even greater attention to unique AGH student city experience of using e-learning techniques in Open and Distance Learning model retaining the abilities of control over the learning process outcomes.

The factor conducive to attainment of that experience was a spatial composition of the AGH campus (see figure 10.1.). The scientific and didactic parts of the university neighbor to compact residential campus (called the Student City) that is a home to approximately 10,000 AGH students. An additional contributing circumstance is the fact that at a half of the distance between research and didactic facilities and residential space of the Student City a supercomputer and networking center can be found called the Cyfronet. That allowed for inclusion of the Student City residents in a broadband Internet gateway.

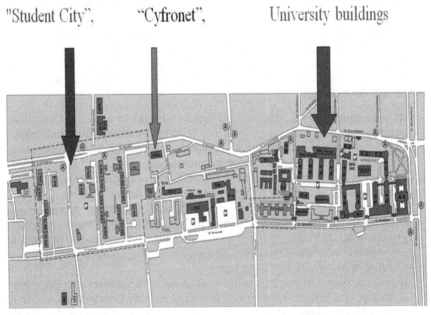

Fig. 10.1 Topography of AGH "Student City", university buildings and "Cyfronet"

10.2 Overview of Open and Distance Learning System in the AGH Student City

10.2.1 A Perspective on the Open Learning in the AGH Student City

As early as in 1998 the author of [5] declared that the AGH university intended to pursue an ambitious scientific experiment aimed to build a model of information-based society using the foundations of the Student City. That was the very first year that the author of this chapter, having been only recently appointed as the rector of the AGH university, immersed himself in a variety of activities leading to transformation of the university of a post-communistic pedigree of "Steel & Coal Academy" (former full official name of the AGH university was University of Mining and Metallurgy and as such it can be observed on figure 10.2.) into modern university with a strong profile in IT (Information Technology). One year later, the English equivalent of the university was successfully changed to University of Science and Technology in which the most dominant faculty accounting for approximately 30 percent of scientific potential was Faculty of Electrical Engineering, Automatics, IT and Electronics. An inherent part of that policy was presentation and offering a broadband and free of charge Internet access (may I underline the importance of the fact that after all, it was merely 1998!) to all of the members of the student community established in the Student City. The community comprised of 10,000 students and PhD candidates and included a group of scientific professionals - AGH University employees - since the city contained facilities such as hotels for individuals employed for assistant positions.

By creating unlimited Internet access in the Student City, the university agreed with the local community council that the inhabitants would accept certain proceedings of scientific observations and research activities enabling the university professionals to closely examine all the phenomenon's related to the process of transformation into post-industrial society (i.e., Information-based Society) on a large scale. The university workers and students of the newly created Faculty of Applied Social Sciences participated in the research leading to identify perform quantitative analysis of technical conditions, fundamental relationships and social and psychological phenomenon's related the aforementioned transformation. The results were considered highly valuable since most of the observed phenomena and processes were formally recorded and analyzed in Poland for the very first time. For example, a survey of the youth members of the AGH Student City allowed for first-time quantitative analysis and characterization of Internet-related forms of addiction [1].

In addition, a preceding experiment of similarly great scale, recorded in the World Guinness Book of Records under a category of "Most Wired Community" designated as the Blacksburg Electronic Village (BEV) was held in the U.S. city of Blacksburg, Virginia. An attempt was made to connect all of the community members altogether (in total 24,000 individuals) [2]. The outcome of that undertaking was relayed through numerous publishers and the scientists specializing in the Internet-related issues. They asserted that both the amount and scientific value of the information that had been accumulated in the Blacksburg experiment was comparable in volume with the information provided at that time by the Mars Pathfinder robot sent by NASA to explore Mars' surface and transfer information concerning its physical and mineral composition.

10.2.2 Evolution of Open and Distance Learning Systems

Thanks to the convenient topographical arrangement of the open distance learning content providers (role of AGH workers of research and didactic backgrounds operating in the facilities can be seen on the right side of the figure 10.1.), network access providers (Cyfronet), the recipients and users of the didactic materials (the students inhabiting the Student City), it was possible to create a university-based network containing a diversity of resources committed to realization of the Open and Distance Learning concept. At first, the resources were made accessible through an additional tab on the university main website (see figure 10.2.). It is worth mentioning that the website was equally accessible to everybody, hence the access to didactic materials was truly unrestricted from the very beginning.

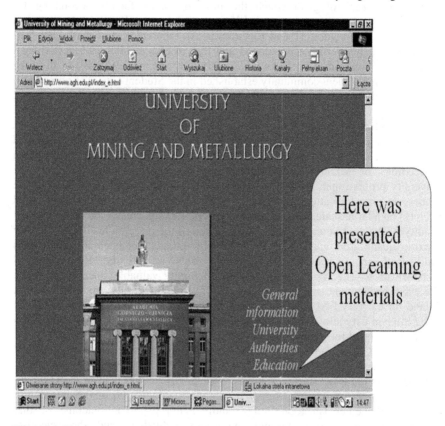

Fig. 10.2 The earliest available forms of providing access to Open Learning didactic materials on the main AGH website

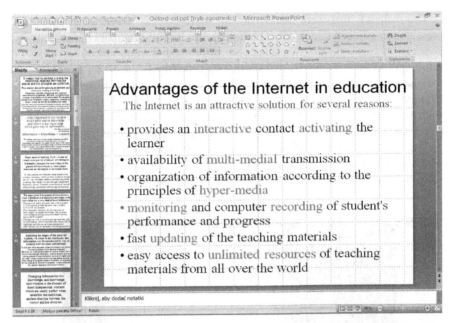

Fig. 10.3. PowerPoint presentation files have been an integral part of the didactic offer contained in the Open and Distance Learning System

Fig. 10.4 The video materials documenting full-length materials relayed through the Open Learning server

The materials that were made available were diverse. Mainly they contained presentation didactic contents prepared in forms originally provided by PowerPoint (see figure 10.3.). What's worth of underscoring, video recordings of full-length lectures were provided starting day one (see figure 10.4.) as were video footage illustrating conducts of experiments performed in the university laboratories (see figure 10.5.). Naturally that had predictable implications for students being able to familiarize with the material during training classes and had a preview of the research activities offered during ensuing laboratory courses.

The Open Learning AGH space offer has also included simulation tools using which a student could perform certain experiments individually using modeling software oriented at technological aggregates (see figure 10.6.).

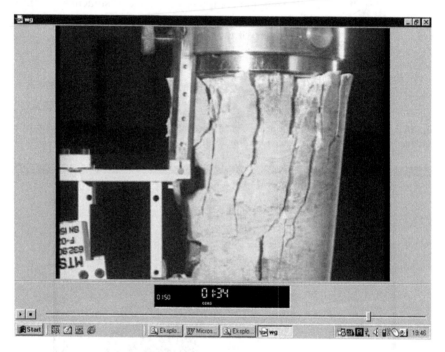

Fig. 10.5 Earlier presentation of laboratory experiment through the Open Learning space improves real conduct of the experiments during ensuing classes

The resources containing reviews and other forms of assessment of educational materials accessible through other internet servers were vital elements of the Open Learning AGH (see figure 10.7.). That kind of tutorial materials will definitely play an important role in development of the Open Learning concept because - speaking from technical standpoint - thanks to the Internet accessing various information resources is nowadays virtually unrestricted. However, that is also met with certain perils because most of the publicly available resources are not formally reviewed, nor even diligently, or accurately prepared. That all leads to a situation resulting in a scenario where web servers on the one hand provide knowledge resources of acclaimed quality, but on the other some of them provide knowledge that is highly imprecise, inaccurate, or falsified. A student (of elementary or higher education institution) seeking information of particular interest is quite unlikely to be able either to verify or to judge whether the knowledge one is being exposed to is noteworthy and sound, or if it is made of deceptive and delusive information. The latter should not be memorized in order to prevent an individual from disinformation which in itself appears to be far more worse perspective than having no information available whatsoever.

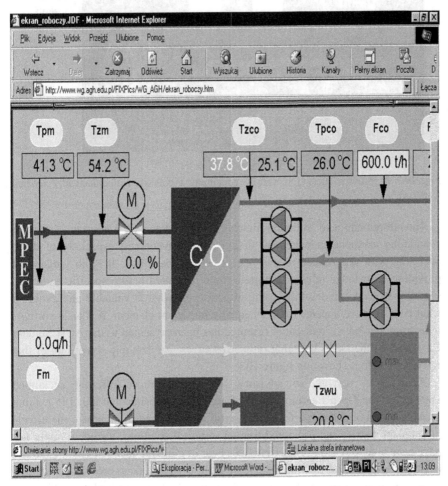

Fig. 10.6 Tools enabling an individual conduct of simulation experiments by students have been available since the beginning of the Open Learning AGH educational offer

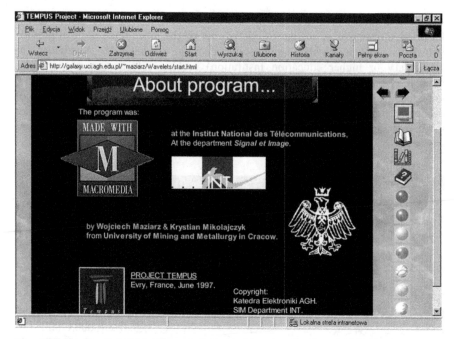

Fig. 10.7 A resource attestation website containing a collection of didactically verified resources

The aforementioned situation indicates an increasingly important factor in conducting adequate management of educational resources, especially of those operating with accordance to the Open Learning paradigm. It is no longer just providing knowledge related to certain topic presented before specific audience. Creating valuable guides facilitating discovery of what is valuable and avoiding what is destructive appears to be an equally important element. It is also worth of indicating that such a referential resource has been presented to the AGH Student City users of the Open Learning online resources. An example screenshot of that reference guide is shown on figure 10.7.

Apart from serving an advisory role on resources offering acclaimed knowledge, the Open Learning AGH offered local websites that aided in gaining access to foreign counterparts. In that sense they allowed to minimize the effects of language barrier and promoted foreign technical solutions, a student might not have been able to reach if otherwise (let's bear in mind the fact that Google did not exist in the current form). An example of such service de facto providing materials of eternal origin is shown on figure 10.8. A lesson of physics is shown offered in Polish which integral part is a streamed (and fully legal) computer animation from one of American educational servers.

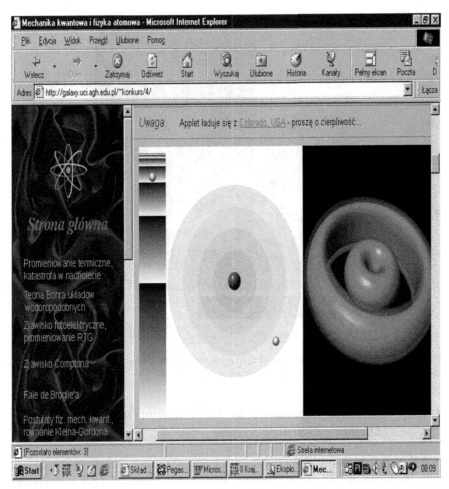

Fig. 10.8 A instructional web page based on external resources. The description is available in the text above

It is also worth of considering the fact that apart from its strictly educational character, the Open Learning AGH operating in conjunction with the Student City AGH played also a role of a communication hub. As in example shown on figure 10.9. includes a snapshot of the Rector's address to candidates and first-year students.

Fig. 10.9 An example of application of the Open and Distance Learning System application aimed to relay announcements, and broadcast to students.

10.2.3   Students' Reactions to Introduction of the Open and Distance Learning System

A first discernible indication of students accepting the educational offer accordant with the Open and Distance Learning System was identified after virtually all student dormitory rooms were equipped with computers with connection to Internet (see figure 10.10.). It should be also mentioned that the hardware was students' a response in form of active private contribution to the offer as the university provided the network and educational contents. The Open Learning educational offer was accessible through computer network on 24/7 basis, so that every AGH university student could access it at his/her discretion regardless of time, or place and choosing a preferable time for studying according to own preferences and predispositions. A result of this greatly computer-aided - thus of strongly individualized and self-reliant character - educational service was a significant improvement on the quality of the educational process.

The positive effect on quality is illustrated by figure 10.11. presenting students' grade average of individuals assigned to a single exercise group that in one point in time in the educational process were granted access to the Open and Distance Learning System and started to benefit from computer-assisted learning. The increase in efficiency of learning can be exemplified by more substantial increases

Fig. 10.10 Computers in students' dormitory rooms as their response to the Open Learning offer made by the university

in student's capacity of acquiring knowledge between verification tests (scores defined on a point scale relating to the subsequent tests are marked on the chart with square dots).

That objectively verifiable phenomenon was confirmed by opinions of students. The Student City community was undergoing a variety of research activities primarily based on opinion polls. Among the inquired information presented to over 3000 students was the question of their assessment of impact the access to the Open and Distance Learning System made on their learning effectiveness. The distribution of results has been presented in a form of chart on the figure 10.12.

In addition it is worth referring to another polling result coming from the same survey. It indicated a volume of knowledge required for taking an exam of introductory course to mathematics (averaged on the basis of entire surveyed population), virtually the same in nature across entire university, that was acquired from the Open and Distance Learning System of particular AGH faculties. Polish designations of the courses shown on figure 10.13. will neither be translated, nor explained since they their particulars are not relevant to this discourse. The aforementioned problem is not related to the fact that the students of the electrical engineering faculty had four times more the access to e-learning in comparison with students of geology, or mining faculties. Actually, it is about a strong diversity in patterns of using the Open and Distance Learning System in comparable learning scenario (students learned the same subject) under identical technical specifics (i.e., 24/7 access to Internet). More details on that matter can be found in [6].

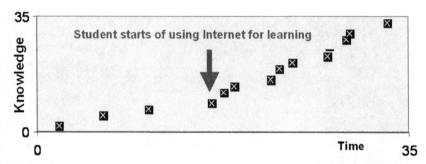

Fig. 10.11 The recorded increase in quality of learning after the introduction of the Open Learning

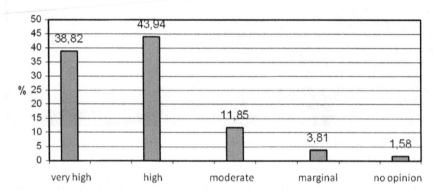

Fig. 10.12 Distribution of the student opinions on the usability degree of the Open and Distance Learning System

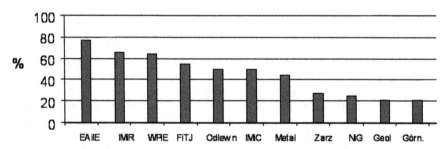

Fig. 10.13 Averaged volume of knowledge acquired from Open and Distance Learning System declared by the surveyed groups of students studying at different AGH faculties

10.3 Continuation of the Open Learning Methodology at the Present-Day AGH University

The Open Leaning idea was originally somewhat exotic in the 1990s since the during creation of the aforementioned system and quoted research. At that time, rough form of concepts, methods of knowledge distribution and majority of published materials were more concerned with theory. The AGH University enabled that idea to materialize using specific conditions provided by the aforementioned university's topography. What's more, that very feature compared to other Polish (and foreign too) universities appeared to be a head of its time.

Nowadays, Open Learning is a renowned and widely accepted system to a degree that its rules are regulated by a ministerial guide [4]. Naturally, the scientists and academics of AGH university could not put down that method of learning and as a result several generations of Open Learning services have

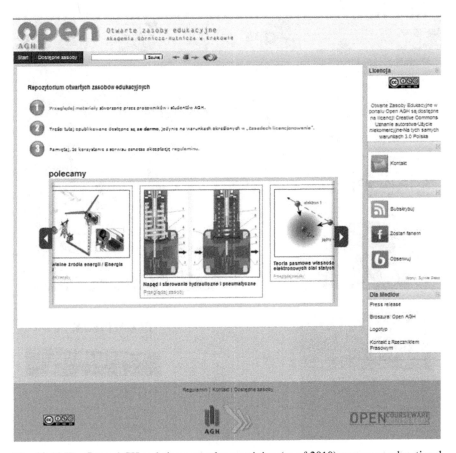

Fig. 10.14 The Open AGH website currently containing (as of 2010) numerous educational resources on the basis of the Open Learning approach

followed since, among which the most recent installment is the Open AGH service. The website in its actual form is shown on figure 10.14. Currently, the website is available only in Polish, hence not all of the presented features may be transparent and comprehendible for the reader. Nonetheless, it's worth showing the details in order to relate further contemplations to this particular specific example.

A direct connection between the Open AGH and acclaimed OpenCourseWare Consortium (OCWC) educational portal is maintained regularly (see figure 10.15.). The portal includes many distinguished members, such as Massachusetts Institute of Technology, Open University UK, UC Berkley and others.

The mentioned direct connection provides the AGH students with access to a variety of very content-rich educational resources virtually originating from all over the world. The subject of currently conducted research activities concerned with   verifying once more how they use the abilities and to what result. As of now

Fig. 10.15 The OpenCourseWare Consortium website is straight accessible through the Open AGH system

there are 4,663 active users and 415 courses are being facilitated in 2009/2010 academic year. The research and observations continue to run and their outcome will be analyzed after collecting a more extensive data once it has been processed.

The Open AGH system is solely based on open source components: the Moodle platform, an ePortfolio Mahara-class system and respective tools originating from the Wikimedia Commons (especially MediaWiki). Both the students and university workers have a blogging system at their disposal (WordPress MultiUser) and special version eduHUB enabling a user to be provided with instantaneous access to all of the modules and their respective resources just after a single log in.

10.4 Copyrighted Materials for Computer-Aided Network-Based Learning Operating in the Open Learning Model

One of the barriers the Open Learning-based education system encounters is the issue of providing access to didactic materials which creation usually entails extreme effort especially as their distribution should, by definition, be universal and free of charge. How should one access such materials?

Firstly, it is reasonable to claim that capabilities provided by computer technology, the Internet network in particular, to the Open Learning-based education should not be limited to exclusive exploitation of ready-made tutoring software. Virtual learning, especially in such noted higher education institution as the AGH university, should resort to use of copyrighted didactic material prepared and modified by recognized experts assuming roles of tutors in respective areas of instruction. Only then such the aforementioned software will be considered useful and applied as assistive, and later on considered a primary tool for conducting particular specializations. Let us investigate this aspect in detail since it is deserves much attention.

Taking into consideration all of the aforementioned qualities of the didactic software distributed to students primarily through the Internet, has a major downside resulting from the fact that it is adapted by software producers to needs of the mass market and more casual users. Yet higher education retains more refined character. In that sense, it is possible to use software, also in a university environment, assisting language learning or the foundations of advanced mathematics. However, no expectations should be derived for similar software offered for tutoring advanced concepts of mechanics, electrical engineering, or thermodynamics. Even if such software were to emerge and available, it is highly unlikely it would match the qualities of regular lecture, or even assistive exercise classes conducted on the level of higher education institution.

Moreover, a few causes for unsatisfactory offerings of such software can be identified in the abundant market of didactic (educational) software. The first and apparently the main cause is unprofitability of software development process (from a software producer's perspective) of such technologically advanced tutoring software. Preparation of that kind of software - even if there were a producer capable and willing to develop it - would prove cost-prohibitive. It would be due to the requirement of highly qualified workforce, and above all the final product would find a moderate group of end users precluding profitability of the development process. What's more, the range and method of academic tutoring concerning monographic subjects of specialization, or skilled trades are subject to constant evolution, development and changes proportionally to the amount of new knowledge absorbed by a faculty. That knowledge results from contemporary studies of increasingly rich science literature and also from the mounting number of original scientific outcomes (produced by tutoring professors) being transposed from the area of scientific research to the didactic process. Therefore, a piece of software equally capable as a modern form of academic lecture conducted over a course of study of any chosen specialization in the AGH university would be subject to frequent updates and content modification. It is also to be mentioned that such lecture is a copyrighted material unique both in its contents and form. Those both factors could not be substituted by a uniform computer software.

The aforementioned causes necessitate an approach aimed to assist Internet-aided learning and application of Open Learning in specialization and major courses delivered during senior academic years of the AGH university, focused on preparation of special didactic materials developed by the best academic teachers. Undoubtedly, that kind of process is laborious, but considered to be definitely profitable, especially in the prospect of network-based distribution.

As a result of consequent application of the Open Learning approach over a course of time and on national scale, Internet servers will inevitably store full texts of various academic lectures. That situation in the context of unrestricted Internet-based access to various resources regardless of their physical location will be conducive to a greater freedom of choice of different interpretations and distinct compendiums of the same topic. Consequently, learning individuals will benefit from a possibility of unconstrained selection of adequate content form stipulated by curricula program that is best suited to their individual preferences. In addition, thanks to abilities of statistical tracking of user visits on a web page, the tutors will be given the information on the interpretations and compendiums that are more preferred by the students. That can hopefully (due to imperfect qualities of human nature) lead to systematic increase in substance and appeal of the forms of published didactic materials. In that context it is certainly reasonable to claim that such form of direct (and by all means confidential - as mentioned before - anonymity of users requesting resources from a web server can be technically facilitated) "popularity contest" of the didactic form in question will define a new quality in the domain of intercommunication between students and teachers. That can definitely lead to increase in the quality of didactical communication. It will also stipulate an emergence of new concept: virtual study script, or even an e-book.

The key problem we need to face with is the issue of obligatory review of didactic materials published on web pages that are being suggested to students. One must not allow the situation of publishing materials serving as tutorials or scientific guide books containing errors and improbable, methodologically speaking, interpretations. Unfortunately, technical flexibility of publishing certain "samizdat" materials (usually coupled with appealing graphical layout) in the Internet leads to real excessive multitude of not entirely sound and true, deceitful, or event totally flawed and highly pernicious from a didactic point of view. Hence, seeing the potential of Open Learning for becoming a major educational opportunity, we need to be aware of its potential of hard to be overestimated danger.

10.5 Conclusions

Similarly to the 1990s, current pattern of activities of the individuals using the Open Learning resources are subject of detailed observation. Furthermore, an extensive research activities are carried out oriented towards determining the consequences of resources use. The need for such research activities appears to be beyond any question. Virtually all of the experts are unanimous in their assessment of current events that potentially lead to further development and popularization of the Open Learning-class systems and as such they appear to be a viable and effective alternative to other forms and methods of public education, especially in the context of the Lifelong Learning initiative and the declared will to establish information-based society, equally called a society of globally accessible information. Unfortunately, when it comes to determining the details concerning the type of learning used in the process of Open Learning and to what effects is still missing essential information.

In addition, a degree of such vagueness practically disqualifies a possibility of using theoretical models in research activities since several assumptions serving as their foundations and confronted with such relatively novel phenomenon may prove to be problematic and highly uncertain. Prognostic usefulness of mathematical methods (i.e., of statistical origin) appear to be similarly doubtful, what also might be the case for computer simulation. It's primarily because each of those approaches exploits the notion of object's current state and certain form of prolongation of currently observed trends. However, we may reasonably assume that the transformation process of contemporary institutionalized model of education will be of bifurcate nature breaking (despite upholding the conditions for continuity in the Cauchy's terms) the principle of smooth transformation. As we know, formal description of such phenomenon's and processes can be achieved through mathematics offering sophisticated apparatus associated with the Benoit Mandelbrott's chaos theory or with the Rene Thom's rapid perturbations in qualitative differential spectra, also known as the catastrophe theory.

Over the time both theories have proven extremely useful, but mainly when they are applied during scientific dissertation and when a science junior seeks an ambitious scientific topic. However, they appear to be totally inapplicable when confronted with an attempt of forecasting of phenomenon's and processes associated with the transformation (on global scale) to use of the Open Learning approach. Hence, regardless of theoretical conceptions pending development in that context, practical experiments are essential in small, select and reasonably well controlled groups of people. As it has been shown in previous chapters, the AGH university has both the capacity and scientific record of administering and conducting such types of experiments. Therefore, such scientific activities should be continued.

Despite its connection with a single university the pursued scientific experiment is likely to enable us to address many questions of detailed nature and its outcomes may prove useful to numerous scientific entities and the economy. This opinion results from the specifics of the AGH university as a very large and highly diverse higher education institution. A background for that diversity can for example be a result of scientific research presented on figure 10.13. We provide here a set of quantitative estimations for more detailed characterization of a degree of that diversification.

Students living in the Student City are taught at the AGH university on a base of 26 specializations. The university offers both types of specializations, the highly theoretical and formalized such as Mathematics, or Physics, and the practical ones such as Mining Engineering, Geology, or Foundry Engineering. In addition, the university provides specializations that are strongly technology-oriented such as Metals Engineering or Drilling of Oil and Gas and also the humanistic counterparts, namely Management, or Environment Protection. The diversity could be easily demonstrated in other aspects of university educational offer including studies of less complex makeup and the ones of more refined and sophisticated composition. It may be also proven using the distinction between studies highly popular among candidates and because of that very quality attracting great numbers of highly capable and motivated attendants. Conversely, there are also specialization studies of fairly decreased popularity, and hence attracting less capable and less motivated individuals. However, further discussion of this aspect is not required and therefore we can proceed to formulate the thesis asserting that the ICT-based (ICT stands for information and telecommunication technologies) research experiment conducted on the population of AGH university students will provide data of quite general quality resulting from strong diversity of that population. It will significantly increase the scientific value of the accumulated observation results and its practical usefulness. It will be because the type of goals and challenges the AGH faculty will face are very diverse. The observation outcomes could also serve in many other universities as an inspiration, or as an optional warning to others.

That higher degree of generality and universality of the accumulated experiment results will likely impact their applicability in the context of various educational models because the structure of didactic process in the AGH university is also quite complex organizationally. It is suffice to refer to 14 faculties among which some offer just a single specialization (e.g. Faculty of Energy and Fuels), whereas some offer a few (e.g. four specializations in information science, automatics and robotics, electronics and telecommunication and electrical engineering offered by Faculty of Electrical Engineering). In addition, the same type of specializations are often offered by distinct faculties (e.g. the Mining Engineering and Geology specialization is offered by five faculties: Faculty of Mining Engineering, Mechanical Engineering and Robotics, Geology, Geophysics and Environment Protection, Mining Surveying and Environmental Engineering and by Faculty of Drilling, Oil and Gas). The number of offered major courses by particular specializations is also diverse. For example, Mathematics, Information Science, or Management and Marketing are administered without further major specialization, but Technical Physics contains six major specializations: nuclear physics, solid state physics, computer physics, medical physics and dosimetry and environmental physics and power engineering.

The aforementioned facts (only a portion of information that could be discussed) have been used to indicate the intrinsic complexity and therefore challenging problem an organization of didactic process is, especially in such large and strongly diverse higher education institution such as the AGH university. It also serves another purpose which is making aware the fact that the range of the scientific experiment discussed in this article can be considered equal to a series of experiments conducted in a selection of higher education institutions of strongly diverse qualities. Many distinct didactic needs will become subjects of research, assessment and analysis in a variety of contexts and despite the experiment conducted in a single university, its outcomes could be used by many educational institutions of highly diverse structure.

References

1. Augustynek, A.: The personal factors for Internet behaviors. In: Haber, L. (ed.) The Information micro-society: based on the AGH Student City, pp. 63–90. AGH Publisher House, Kraków (2001) (in Polish)
2. Cohill, A.M.: Education versus Technology: The Evolution of the Blacksburg Electronic illage, CNI (1995),
   http://www.cni.org/tfms/1995b.fall/BEV.html
3. Haber, L.H. (ed.): The Information micro-society. AGH Publisher House, Kraków (2001) (in Polish)
4. Ministry of Science and Higher Education: Open Learning Guide (2010)
5. Tadeusiewicz, R.: The information society model. Forum Akademickie 12, 28–30 (1998) (in Polish)
6. Tadeusiewicz, R.: The application of CAL methods for Information Society development. In: 8th Polish Conference Information Technology in Education Process, Kraków, pp. 57–75 (1998) (in Polish)